5th DEC '87

Ben,

Best wi

speedy ...

Tim

Computer Simulation
in
Physical Geography

Computer Simulation in Physical Geography

M. J. Kirkby

P. S. Naden

T. P. Burt

D. P. Butcher

JOHN WILEY & SONS

Chichester · New York · Brisbane · Toronto · Singapore

Copyright © 1987 by John Wiley & Sons Ltd.

Library of Congress Cataloging-in-Publication Data:

Computer simulation in physical geography.

   1. Physical geography—Mathematical models.
2. Physical geography—Data processing. I. Kirkby, M. J.
GB21.5.M33C65   1987   910′.02′0724   87–8125

ISBN 0 471 90604 2

British Library Cataloguing in Publication Data:

Computer simulation in physical geography.
   1. Physical geography—Simulation methods 2. Physical geography—Data processing
   I. Kirkby, M. J.
   910.02′0724   GB21.5.S55

ISBN 0 471 90604 2

Photo-typeset by Photo-Graphics, Honiton, Devon, Great Britain

Printed and bound in Great Britain

Contents

Preface

During the 1980s computers have changed from being typically large main-frame machines for skilled users at large centres, to being most commonly seen in the form of desktop microcomputers which are accessible to a much larger number of people who have come to be aware of them in the home, at school or in the office. In this book we have tried to lead some of these new users with interests in physical geography towards the use and programming of simulation models. These models may be used in sixth form and university courses both to learn about aspects of the real world in an experimental situation, and as a relevant context in which to learn programming. The material in this book is developed from a workshop course given in September 1985 as part of the First International Geomorphology Conference, based in Manchester.

In any book which includes computer program listings, there is a difficulty in achieving any degree of transferability between machines. This is particularly true for the many dialects of the programming language BASIC, which is overwhelmingly the most common in use on microcomputers and has, therefore, been adopted for use here. Some degree of transferability has, however, been achieved by writing the programs in the main part of the text using only a standard subset of commands which are almost, though not completely, common to all common microcomputers. The result of this is to make the programs rather less succinct and elegant than they might have been if they had taken full advantage of the particular features of any single machine. Where machines are known to differ from the standards used, some comments have been given for guidance in the relevant appendix. The restrictions imposed by these standard commands are most severe with regard to input and output, both for text and graphics. The solution adopted has been to use standard line numbering to refer to a series of input and output subroutines for formatting text and drawing simple graphs and histograms on scaled axes. These subroutines are specific to individual types of microcomputer, so that they need to be rewritten for each machine. Appendices specify the function of each subroutine, and include listings for Acorn 'BBC' in BBC Basic, and for IBM PCs in BASICA. Users of other systems may be able to use these listings as a basis for adaptation. The listings in the text, together with the appropriate subroutines for these machines, are also available on disk, and may be ordered from the publisher in an appropriate format.

The book is divided into two parts. In Part 1, after an introductory chapter on our view of models in physical geography, and introducing the program and subroutine structure used, Chapters 2 to 5 present a series of simulation models which are grouped into model types. Chapter 2 is concerned with 'Black Box' or input–output models; Chapter 3 examines process models; Chapter 4 deals with mass balance models and Chapter 5 with stochastic models. It is recognized that these types of model are not mutually exclusive, and that many models contain elements of one or more types. Within each chapter individual models are placed primarily in their geographical context, to emphasize the range of problems to which related simulation models may be applied, and to allow the models to be used with a minimum of prior computer expertise.

In Part 2 the focus of the book shifts from the geographical context to the computing rationale. Chapter 6 gives advice on the construction of a computer simulation model from the stage of problem definition to the completion of a finished and working program. Chapter 7 is concerned with problems of choosing appropriate parameters, both in terms of optimization of a forecasting model to match an observed outcome, and in terms of sensitivity of the outcome to parameter changes. Chapter 8 addresses the issue of defining one's problem, illustrating the range of alternative model types which may be appropriate to a particular geographical context and guiding the reader towards the difficult initial step of seeing a problem in a way which is amenable to simulation. The book is completed with a series of appendices describing the standard set of BASIC commands used, the subroutine specifications, and subroutine listings for BBC and IBM microcomputers.

In the world of computing, technological advance has a habit of overtaking publication but it is hoped that the principles set out here will be of use for several years. Even if the BBC microcomputers on which the programs were developed have become obsolete, the geographical context will remain relevant to a growing number of users in both schools and institutions of higher education, as more and more people discover that computer simulation is a cost-effective tool for extending and understanding our field observations and, hence, learning about our natural environment.

March 1987

Mike Kirkby
Pam Naden
Tim Burt
Dave Butcher

Part I

CHAPTER 1

Introduction: Getting Started

1.1 WHAT IS A MODEL IN PHYSICAL GEOGRAPHY?

A model simulates the effect of an actual or hypothetical set of processes, and forecasts one or more possible outcomes. At one extreme the simulation may be as simple as substituting values into an equation or matching a pattern. At the other extreme the simulation may closely follow the detailed processes which operate in the real world. Most current models are deterministic, so that for a given set of inputs there is a unique forecast. Others are stochastic: that is they contain at least some random or chance element in the process operation or the inputs to the model, so that more than one, and usually a very large number of outcomes are possible.

Models can never fully represent the real world, but can only be analogies or analogues which have some features and behaviours in common with it. They differ very widely in their degree of similarity to the real world prototype, and in a number of ways. For example they differ in how much they physically 'look like' the prototype and in how well they forecast its behaviour.

Physical resemblance does not necessarily guarantee that a model is effective. To say that the moon looks like green cheese tells us very little about its true properties! On the other hand accurate scale models are sometimes used to model landform features and forecast their response, and this approach has been successfully used to forecast, for example, coastal erosion and sedimentation, river channel changes and drainage basin erosion. Physical similarity does not necessarily require an obvious physical resemblance, but must depend on the operation of similar physical laws in the model and real world. For example the (laminar) flow of water in a saturated aquifer is proportional to the hydraulic gradient and the hydraulic conductivity of the aquifer material (Darcy's law). Similarly flow of electricity (current) is proportional to electrical pressure (voltage) gradient and electrical conductivity (1/resistance), through Ohm's law. Thus flow through an aquifer may be modelled by measuring current in a conducting material of the same physical shape as the aquifer. This kind of analogue model is one step along a road towards an abstract representation of each process in the real world by a mathematical or logical expression. Thus the rate of flow in the aquifer is represented or modelled by the mathematical expression:

$$Q = KG \qquad (1.1)$$

where Q is the flow through each square metre of aquifer
K is the hydraulic conductivity of the aquifer
G is the hydraulic pressure gradient.

In this form the flow can be calculated at each point, either with a pencil and paper, or using a calculator or computer. If a computer model calculates flow from this expression, the model may still be said to resemble the real world, even though the computer does not physically 'look like' an aquifer.

It is clear that computer models of environmental forms or processes can only look like the real world at this abstract, mathematical level. This kind of similarity does nothing to confirm or refute the effectiveness of a model to produce realistic forecasts. It is easy, and plainly wrong, to assume that because the computer model relies on abstract mathematical expressions it must be right in its forecasts. It is equally easy, and equally wrong, to assume that because the computer model has no physical resemblance to a landscape then its forecasts must be incorrect. The truth generally lies between these two extremes for computer models, as for all kinds of models, hypotheses or theories. No model is better than the assumptions and data it relies on. In circumstances where the assumptions are valid and/or forecasts are being made within the range of data used to establish the models, then it will probably give reasonable results. Where one is forecasting outside this area of tested reliability, no model should be relied on without further testing. In other words the well-worn computer dictum of 'Garbage in – Garbage out' applies to models as to other computer operations, and the inscrutability of the computer should never be confused with reliability.

If a model is to be of any use, either practical or theoretical, it should produce testable outcomes. Computer models are necessarily numerical and logical, so that they are one form of quantitative model, producing a definite forecast in each run of the model. In many cases this forecast can be compared numerically with real world measurements (which should be independent of the data used to set up the model) to make a very exact test of performance. It is, however, also possible to make qualitative deductions from model output which can be compared with the real world, at a less precise although not always at a 'worse' level. In fact, where models have a stochastic element, either in the processes they represent or through uncertainty in the constants put into them, the numerical outcome must become less precise, so that there is a continuous range of models from strictly deterministic quantitative models with a unique outcome; through a range of stochastic models in which outcomes are more or less probable; to entirely qualitative models including many of those which have been traditionally successful in physical geography like W. M. Davis' geographical cycle of erosion.

Models which are suitable for computer programming must be expressible in the form of a strictly logical and/or numerical procedure. Most of the examples in this book are mainly deterministic though some have a significant random component. A logical procedure is necessary if a model is to be implemented on a computer, but not all logical models make useful programs. In some cases the trouble taken in writing a program does not justify the time saved in following the procedure, however logical. This is particularly true if the model will only be used a few times.

1.2 TYPES OF COMPUTER MODEL

The various types of computer model are discussed in Chapters 2 to 5. Although they build up in some senses from simple to complex, the categories overlap considerably, and many working models contain elements of several types. The categories used here are 'black box models' (Chapter 2), 'process models' (Chapter 3), 'mass balance models' (Chapter 4) and 'stochastic models' (Chapter 5), although there is no agreed categorization of model types along these or other lines. In the simplest models, the workings of the model are treated as invisible to the user, who feeds in input and takes out forecasts. Most computer models can be treated in this way by their users, even though we hope that you will not be content with this level of knowledge! Such models are called 'black box' or 'input/output' models. These terms are usually applied to models in which the internal workings of the model are not intended to directly represent the processes operating in the real world, even at an abstract mathematical level. Perhaps the most widespread example of this type of model is where input and output are related to one another by a statistical curve fitting or other regression procedure. Output may then be estimated from input data, but little is learned about the operation of the linking process. This type of model may be very effective, and in cases where processes are poorly understood is likely to be the best available in terms of forecasting accuracy. It tends, however, to behave worse than process and/or mass balance models when input lies outside the range of the original test data.

Both process models and mass balance models attempt to shine some light into the black box. If all processes and relationships are fully represented in the model, then we may be said to have a 'white box' model. Such a model is usually exceedingly complicated, even where we have the knowledge to construct it, so that most achievable models may be considered as boxes in various rather dark shades of grey. There is always a conflict between greater reality, associated with greater complexity, and greater simplicity at the expense of detail in representing reality. The 'best' model can only be judged by how well it satisfies its original purpose.

Process models describe the mechanisms of particular operations which occur in the real world. For example a black box model of soil erosion may estimate erosion rates in terms of empirical equations, from storm rainfall, slope length and slope gradient. In a process model the erosion may be separated into rainsplash, and sheetwash in overland flow. Rainsplash may be forecast in terms of rainfall intensity and soil properties; and sheetwash through estimates of overland flow and its transporting capacity. In other words the model is built up from a flow diagram which represents the physical storages and/or flows of energy or material in the real world. The process model can never be perfect or complete, but it is an attempt to make the conceptual behaviour of the model resemble the real world more closely than in a black box model. In many if not most models there are many processes operating at once, and interacting with one another. The overall model can therefore be assembled from a number of sub-models, each representing a single process or group of processes. In an effective model the various sub-models need to be dealing in flows of the same commodity, usually mass or energy; and to be dealing in it at similar levels of space and time resolution. It is, for example, extremely difficult to reconcile a model of atmospheric circulation which deals in square 100×100 kilometre 'cells' with a hydrological model which works with irregularly shaped small drainage basins of about one square kilometre.

A very important set of constraints on most models is that, barring nuclear explosions and radioactive decay, mass and energy are neither created nor destroyed. This conservation applies not only to total mass, but to the masses of individual chemical elements, like, say, iron or carbon. It also applies to compound materials, provided allowance is made for chemical changes and change of state. Perhaps the most obvious example of this principle, and one of the most important for physical geography, is the hydrological cycle. It requires that the mass of water is conserved if due allowance is made for chemical changes like release from volcanoes or incorporation into sediments; and for changes of state to and from ice and water vapour. In a similar way we can generally appeal to the conservation of total rock and soil materials. In following the course of weathering or nutrient cycling we may be equally interested in budgeting say silicon or carbon. In all of these mass budgets there is little loss of material from the system of interest, so that the relevant budgets form very effective overall controls of many of the systems we will be looking at. Although energy is also conserved, energy balances prove somewhat less useful than mass balances in most of the cases we will look at. The reason for this preference for mass balance is that there are large losses of energy in most mechanical systems, and the losses are not well understood, so that many of the largest terms in the energy budget are uncertain. There is, however, one important exception where an effective energy budget is crucial: in modelling the temperature of the earth's surface and evapo-transpiration from it. For this essentially thermodynamic system the energy losses are relatively small.

Wherever mass or energy balances are appropriate, the model is constrained by some form of the 'storage equation':

$$\text{Input} - \text{Output} = \text{Net increase in storage} \qquad (1.2)$$

This equation applies not only to the system as a whole, but to each spatial compartment of it. It is applied most commonly to mass of water or earth materials, and less commonly to mass of individual elements, to total or radiation energy or to biological populations. The importance of the mass balance approach cannot be overstressed. It provides a common strand to many models of interest, and gives many of them a distinct family resemblance, as will be seen in Chapter 4. In opening up the black box model, mass or energy balance may be thought of as providing the framework for a physically based model, within which the individual process models are supported. Storage equations also have an important role in providing the formal link between rates of change of flows across space and rates of change of state over time at a point. In equation (1.2), the flows provide the input and output terms on the left-hand side, while the change of state is the increase in storage on the right-hand side. A final and very practical virtue of mass or energy storage equations is that they have a much better physical foundation than many of the processes which we hang around them, so that they help to keep forecasts within the range of the possible. In fact it may be argued that in some large models much of their overall forecasting power is based on the constraining effect of mass balance rather than on our reliable understanding of the processes acting.

The last type of model discussed in Part I is the stochastic model (Chapter 5). This category tends to run across all the others, although the simplest models generally lack a stochastic element. Random elements are usually included to represent processes or forms which are outside the scope of the model. The real world we are attempting to model may usually be conceived as strictly deterministic, so that processes are not random in principle. Nevertheless we may not wish to include the causes of every process within our model. These processes may then be represented by a series of random numbers drawn from a specified probability distribution. Some examples show the sort of cases in which this approach is useful. If we have a hydrological simulation model which converts rainfall into stream runoff, we can use it to forecast the runoff from a specified storm sequence. If, alternatively we want to know the size of the flood which will, on average, be exceeded every 100 years, then a simulated random rainfall sequence may be generated from our knowledge of local rainfall distributions. This sequence may then be used to forecast flood sizes and their distribution, without needing a very long real runoff record. This approach is likely to be more cost effective and more reliable than attempting to forecast rainfall from a global circulation model.

A second example in which a stochastic model may be appropriate is in producing the initial surface on which erosion occurs. This will always have some faint irregular relief, but we are not usually interested in its exact form. We might therefore generate the irregularities as suitable random numbers instead of investigating and modelling the causes of the micro-relief. In each of these two examples, the use of random numbers does not imply that rainfall or relief could take *any* value. Instead the random numbers would be drawn from a very definite probability distribution. A random number might, for example, be drawn from a normal distribution with an average of 100 and a standard deviation of 1, so that most values lie between 98 and 102. Randomness may be constrained to this or any extent, and certainly does not mean that model outcomes lack regularity; but only that outcomes are not uniquely determined.

There is another kind of stochastic variation which is not included in Chapter 5, but which is important in testing models. The outcome of any model is determined by the *parameters* which are used to describe initial states, process rate constants, etc. Most of these parameters are not known to a high degree of accuracy, because of experimental error and field variations from site to site. If each parameter value is drawn from a random distribution which represents its range of possible values, then a series of runs of the model, even of a deterministic model, will give a distribution of outcomes. This distribution may then be compared with field site measurements to determine whether the real world values lie within the range of forecast outcomes of the model. This topic is taken up again in Chapter 7.

1.3 WRITING A COMPUTER SIMULATION MODEL

This section provides a brief introduction to model development, so that you can read and use the programs of Part I. Methods of model creation and development are discussed in Chapter 6, but are briefly summarized in the flow diagram of Figure 1.1. The first step is to define a procedure for forecasting on the basis of your understanding of the environmental problem. This procedure will usually conform to one or more of the model types discussed above, and must ultimately be expressible in logical or numerical form. The best guide to developing suitable procedures is to follow other examples as far as appropriate. However, this is not always possible and you will at times need to refer directly to scientific or mathematical principles.

Once a procedure or algorithm has been designed, the various steps of the routine must be put into a definite order of operations, which the computer must perform. Even though a number of operations may in the real world be simultaneous, the computer must be instructed to do them in a definite order. Construction of a flow diagram like Figure 1.1 is helpful to clarify this sequence. The next step is to ask how you wish to enter the parameter values and extract data on the model outcome at one or more stages. Values may be entered from the keyboard, through DATA statements in the BASIC program, by calculation from a mathematical or logical function, or from a file stored on magnetic disc, tape or other medium. In the simple programs illustrated in this book, the last of these is not usually used, but for models with large data requirements a file is a likely choice. In many cases it is helpful to the program user, even if only yourself, to prompt keyboard entry with a plausible value or range of values, which can then be revised by the user. This method is illustrated below. Output from

```
┌─────────────────────────────────────────────────┐
│ Define procedure in numerical/logical form       │
└─────────────────────────────────────────────────┘
                        ↓
┌─────────────────────────────────────────────────┐
│ Create flow diagram showing steps in sequence     │
└─────────────────────────────────────────────────┘
                        ↓
┌─────────────────────────────────────────────────┐
│ Choose inputs/outputs and their format             │
└─────────────────────────────────────────────────┘
                        ↓
┌─────────────────────────────────────────────────┐
│ Incorporate input/output into flow diagram         │
└─────────────────────────────────────────────────┘
                        ↓
┌─────────────────────────────────────────────────┐
│ Write BASIC program                                │
└─────────────────────────────────────────────────┘
                        ↓
┌─────────────────────────────────────────────────┐
│ Test for program errors and correct                │
└─────────────────────────────────────────────────┘
   Not OK                              OK
┌─────────────────────────────────────────────────┐
│ Test for cases where outcome is known              │
│ and correct  ⎰ procedure                           │
│              ⎱ flow diagram                        │
│                BASIC                               │
└─────────────────────────────────────────────────┘
   Not OK                              OK
┌─────────────────────────────────────────────────┐
│ Refine and modify                                  │
│              ⎰ procedure                           │
│              ⎱ input/output                        │
│                BASIC                               │
└─────────────────────────────────────────────────┘
   Not OK                              OK
              ┌──────────────┐
              │ Use Model    │
              └──────────────┘
```

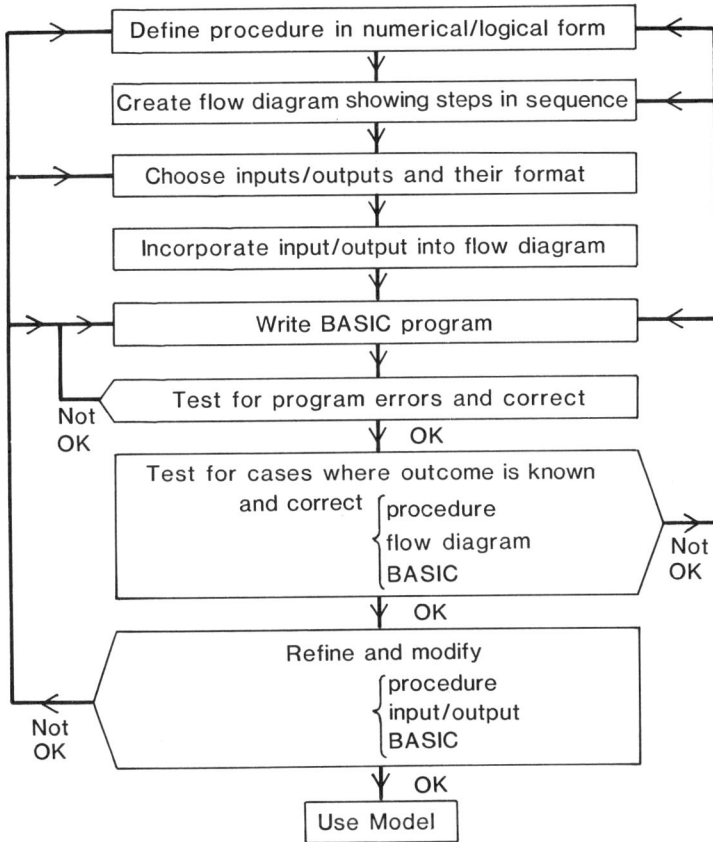

Figure 1.1 Summary procedure for designing and writing a computer model

the program is primarily through the monitor/VDU/TV screen, and may be in text form or as some type of graph (or other graphic design). Where a printer is connected, this screen output may be replicated on paper. Another possibility for large models is to store results to disc or tape as a data file. The flow diagram now needs to be modified to fit input and output into the sequence of computer operations. While developing a program and testing it for errors, it is often useful to obtain output at many more stages than will be required in the final running model.

With what is now a complete flow diagram, you are in a position to write a program in BASIC or any other computer language. If other people are to use your program, and even if you wish to return to it after the lapse of a few days or weeks, it is helpful to sprinkle the program text with remark (REM) statements to explain what the program is doing, preferably in terms of the flow diagram.

The last stages in program development are to test the program at three levels. Firstly, it is necessary to make the program run without producing error messages. This stage becomes shorter with practice but is never eliminated. Once the program runs in some form, it must be tested for cases where the correct outcome is known from the model assumptions. This is usually done for simple cases where you can reach a mathematical solution or work through the procedure by hand calculation. During this stage it is necessary to check that every piece of the model is doing what you think it should. The third stage is to refine the model, and modify flow diagrams and or BASIC accordingly. One type of refinement is to make the model more convenient to use, usually involving changes in input and/or output. You may also wish to refine the model to improve either its speed of operation by eliminating unnecessary steps, or its representation of reality as defined in the original algorithm. This stage of refinement can be very time consuming. It should be terminated as soon as the model meets its objectives: to attempt perfection is unrealistic.

The resumés of this chapter do not constitute a short manual in computer modelling, but are meant to give enough of a glimpse of the various techniques, model types and approaches to tackle the next four chapters. Before you go on, you should run the brief programs set out below in this chapter, and familiarize yourself with the way in which program listings are presented in this book. Our overall aim is to cater for modellers at more than one level, and we expect few to absorb all the material at a first reading. For the newcomer to modelling, Part I provides an undemanding route into the use of models which are related to earth surface problems. At a slightly higher level, Part II provides the necessary background to help you to write simple models for yourself, related to a chosen problem. Our aim is to provide the principles and methods which can be used to develop models which meet real forecasting, teaching and research needs.

1.4 HOW TO USE THE PROGRAM LISTINGS

The program listings in Chapters 1 to 8 have been written in a rather minimal core subset of the BASIC commands available on most current microcomputers. Although all of the programs have been written and tested on a BBC 'B' 32K-byte micro, they can be used with little or no modification on a number of other machines, and it is our intention to concentrate on common rather than machine-specific aspects of BASIC to allow this kind of transferability. Appendix A sets out the BASIC keywords which have been used in the programs, together with some rules for allowable variable names, etc. It also contains some notes on the minor adaptations required to transfer the programs to some other machines, including the IBM PC and compatible machines.

Most microcomputers support text input and output commands which, although always present, vary somewhat in syntax. There are wider divergences between graphics commands, although they too are generally available in some form. In order to make some use of these facilities, while retaining the principle of transferability, the listings make use of a standard set of subroutines, numbered as statements 10000 and above. These subroutines provide formatted input and output of text, and the ability to draw simple graphs and histograms. Both assume a minimum specification which almost all micros can meet, of 40-column text and single-colour graphics. Appendix B provides a listing of these subroutines for the BBC 'B' and IBM PC micros, together with details of exactly what each subroutine does. These programs are each machine specific. To use the listings on another machine, these subroutines need to be replaced by a program which performs the same functions, at the same subroutine statement numbers, for the machine in question. The main listing will then function correctly for the other machine.

Before attempting to modify any of the main listings yourself, you are advised to make yourself familiar with the functions of the various subroutines in PROTO, so that you can use them correctly. Before writing new programs, you will need either to be familiar with PROTO, or with the corresponding graphics and input/output facilities on your target machine. In the long term, the latter course is strongly recommended, because the PROTO routines rarely make the best use of each micro's strengths, or avoid its weaknesses.

As an example of the use of PROTO, three short programs are included here to show the use of the subroutines for handling text and graphics. Figure 1.2 is a short routine, INDATA, for entering a series of parameter values from names and suggested values and limits held in DATA statements. This information is used to cue and check the input values, and only move on when a satisfactory value is entered. Figure 1.3 is a routine, OUTDATA, to demonstrate the output of a table of data. In the example, subroutine 1000 generates a multiplication table, but this would be replaced as required. When the bottom of the screen is reached, output halts to print out the values to paper or until told to proceed. Figure 1.4 is a routine, GRAPH, to draw the graph of a function, and print it out if a printer is connected. You are asked to enter a function and its curve will be drawn over specified ranges of x and y, as is shown for a simple example in Figure 1.5.

Routines of this kind are included in many of the listings later in the book. To understand them fully, you will need to refer to Appendix B, but you should be able to run the programs immediately, and pick up the detail at a later stage, as needed.

```
  10 GOSUB 10000:REM Initialize text screen
  20 RESTORE:DIM X(50):
     PRINT "SIMPLE DATA INPUT ROUTINE":C=0:R=2:I=0
  30 READ C$:
     IF C$="END" OR C$="End" OR C$="end" OR I=50 THEN END
  40 C$="Enter "+C$+": " :READ A,LL,UL
  50 GOSUB 12000:IF A<LL OR A>UL THEN GOTO 50
  60 X(I)=A:R=R+1:I=I+1:IF R<23 THEN GOTO 30
  70 B$=C$:R=23:A$="Press RETURN to continue ":A$="":
     GOSUB 11000
  80 GOSUB 18000:R=0:GOTO 30
1000 DATA Width in m,10,0,100
1010 DATA Depth in m,1,0,5
1020 DATA Velocity in m/s,0.8,0,5
1030 DATA end
```

Figure 1.2 Listing for INDATA

```
  10 GOSUB 10000:REM Initialize text screen
  20 PRINT "SIMPLE DATA PRINT-OUT":R=2:C=0:B=4:DIM Z(8,30)
  30 GOSUB 1000:REM Generate data !
  40 REM Printout begins here
  50 FOR J=1 TO 30
  60   FOR I=1 TO 8
  70     C=5*(I-1)-1:A=Z(I,J):GOSUB 17300
  80     NEXT I
  90   R=R+1:IF R=23 THEN GOSUB 500
 100 NEXT J:END
 110 REM ===============
 490 REM Waits before clearing screen
 500 C=0:C$="Press RETURN to continue ":A$="":GOSUB 11000
 510 GOSUB 18000:R=0:RETURN
 980 REM ===============
 990 REM Values for demonstration
1000 FOR I=1 TO 8
1010   FOR J=1 TO 30
1020     Z(I,J)=I*J
1030     NEXT J
1040   NEXT I
1050 RETURN
1060 REM ===============
```

Figure 1.3 Listing for OUTDATA

```
  10 GOSUB 10000:GOSUB 20000:REM Initialize and open graphics
     screen
  20 XX=8:NX=-XX:XY=5:NY=-XY:GOSUB 21000:REM Set plotting ranges
  30 SX=5:SY=2:GOSUB 22000:REM Draw and scale axes
  40 C=0:R=0:C$="Enter title: ":A$="GRAPH":GOSUB 11000:REM Enter
     A$, cued by C$ at C,R
  50 X=-8:Y=4.5:GOSUB 23000:B$="Y(X)=X*X/5":REM Title on graph
  60 C$="Enter function to plot: ":A$=B$:GOSUB 18000:
     GOSUB 11000:B$=A$:GOSUB 13000
  70 FG=0:FOR X=NX TO XX STEP (XX-NX)/50
  80   Y=FNY(X):GOSUB 26000:NEXT X:REM Plot curve
  90 C$="? Superimpose another graph: (Y/N): ":A$="Y":
     GOSUB 18000:GOSUB 11000
 100 IF A$="Y" OR A$="y" THEN GOTO 60
 110 GOSUB 29500:REM Dump graph to printer
 120 END
```

Figure 1.4 Listing for GRAPH

? Superimpose another graph: (Y/N): N

$y(x) = A*x*x + B$

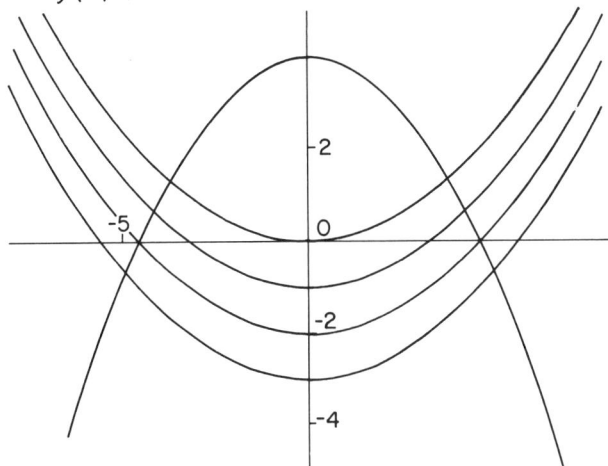

Figure 1.5 Example output from GRAPH for parabolas of the form $y(x)=Ax^2 + B$. The curves shown are for $A = 1/10$; $B=0, -1, -2$ and -3; and for $A = -1/5$; $B=4$

1.5 BACKGROUND READING

1.5.1 Environmental systems

Anderson, M. G. and Burt, T. P. (eds.) (1985) *Hydrological Forecasting*, John Wiley, Chichester.
Carson, M. A. and Kirkby, M. J. (1972) *Hillslope Form and Process*, Cambridge University Press, Cambridge.
Kirkby, M. J. (ed.) (1976) *Hillslope Hydrology*, John Wiley, Chichester.
Kormondy, E. J. (1984) *Concepts of Ecology* (3rd edn), Prentice Hall, New Jersey.
Richards, K. S. (1982) *Rivers: Form and Process in Alluvial Channels*, Methuen, London.
Selby, M. J. (1982) *Hillslope Materials and Processes*. Oxford University Press, Oxford.
Trudgill, S. T. (1977) *Soil and Vegetation Systems*. Oxford University Press, Oxford.

1.5.2 Models of environmental systems

Ahnert, F. (1976) Quantitative slope models, *Zeitschrift fur Geomorphologie*, *Supp. Band.*, **25**.

Harte, J. (1985) *Consider a Spherical Cow.* Wiliam Kaufmann Inc., Los Altos, California.

Jeffers, J. N. R. (1982) *Modelling. Outline Studies in Ecology.* Chapman and Hall, London.

Maynard Smith, J. (1974) *Models in Ecology.* Cambridge University Press, Cambridge.

Thomas, R. W. and Huggett, R. J. (1980) *Modelling in Geography: A Mathematical Approach.* Harper and Row, London.

Woldenberg, M. J. (ed.) (1985) *Models in Geomorphology,* Allen and Unwin, Boston.

1.5.3 Background mathematics and science

Davidson, D. A. (1978) *Science for Physical Geographers.* Edward Arnold, London.

Wilson, A. G. and Kirkby, M. J. (1980) *Mathematics for Geographers and Planners* (2nd edn), Oxford University Press, Oxford.

Chapter 2

Black Box Models

'The cause is hidden but the result is known' — Ovid

2.1 THE STRUCTURE OF SYSTEMS

The real world is complex and continuous. Science aims to isolate simplified functional units from this complexity and to investigate how such 'systems' operate. Chorley and Kennedy (1971) have defined a system as 'a structured set of objects and/or attributes. The objects and attributes consist of components or variables (i.e. phenomena which are free to assume variable magnitudes) that exhibit discernible relationships with one another and operate together as a complex whole, according to some observed pattern'. The science of systems analysis enables the structure and behaviour of systems to be explored. All systems analysis seeks to define functional units which are simple enough to understand and investigate and yet which are sufficiently complex to be of some significance. As implied in Chapter 1, the link between systems analysis and computer modelling is achieved by representing the system as a mathematical expression.

All environmental systems are subjected to input (X) which are transformed by the system (S) to provide output (Y). Thus

$$Y = S(X)$$

The transformation operator S is known as the transfer function and is, in effect, the system itself in that it accounts for the way in which input is translated into output. Since S is not equal to 1 (except in trivial systems), we need to define the nature of the transfer function. This will consist of a mixture of parameters (or constants) and variables which together define the nature of the transfer function. Two approaches to the study of input/output systems may be taken:

1   White box models: The structure of the system (i.e. the transfer function) is built up from knowledge of the variables involved in the system and from an understanding of the relationships between those variables. In Chapters 3 and 4, examples of such models will be given by examining storages and processes operating within the system.

15

2   Black box models: As the name implies, here the system is treated as a single unit without any attempt to unravel its internal structure; the transfer function is taken to have a single, constant value therefore. Rather than being based on theoretical knowledge, black box models are usually defined experimentally (or empirically) and result from 'measurement knowledge'. Such measurements are particularly important in the early stages of any study, since they allow initial identification of the processes involved in a particular system, and the establishment of relationships between the variables. Black box systems are essentially 'functional' therefore, whilst the mass balance and process models described later can be termed 'realistic'. A further category of system, the grey box, has also been recognized: in this a partial view of the structure of the system is achieved. The internal construction of identified subsystems is not considered, however, and in most respects the grey box approach is simply the result of experimental investigation of black box systems rather than representing a real transition towards the definition of a white box system.

Of course, it must be stressed that this classification is somewhat arbitrary: all models are really a shade of 'grey' since even the most realistic models involve some simplification of the real world. Here we concentrate on those models which contain little or no process information.

2.2   REGRESSION MODELS

A morphological system was defined by Chorley and Kennedy (1971) in terms of the degree of association which exists between the physical properties (e.g. geometry, composition) of the system. Morphological systems are defined using regression and correlation techniques and as such represent the least complex of natural systems. The raw data for these statistical analyses consists of paired observations of two variables within the system. One is termed the independent variable (X) which is considered to control the dependent variable (Y). Note that X and Y can refer to the input into, and the output from a system, or to the relationship between two variables from within the system. The identification of X as the independent variable and Y as its dependent implies a cause-and-effect relationship between the two. However, such simple causality rarely exists within an environmental system and great caution is needed to say why a given correlation occurs. Two variables may be covariant for one of several reasons (Chorley and Kennedy, 1971, p. 24):

1   X is the true cause of Y. Such a conclusion often demands a greater knowledge of the system than is usually the case in black box analysis.
2   X and Y are correlated because of the outside influence of one or more truly independent variables. For example peak flood discharge and the

volume of food runoff may be highly correlated, but both are of course controlled by the magnitude of the rainfall input as well as by antecedent catchment conditions.

3   X is the true cause of Y but only indirectly via several intermediate variables. For example soil moisture conditions modify the relationship between rainfall and runoff.

4   A spurious or chance correlation. In this case, an enlarged data set or collection of a new sample should provide a new correlation coefficient with a value much closer to zero.

The association between X and Y may be most easily seen on a scatter diagram (Figure 2.1(a)). The scatter diagram represents a simple model which may be approximated by fitting between X and Y. The linear regression model takes the form

$$Y = a + bX$$

The intercept (a) and the slope of the regression line (b) are the model parameters. The sign associated with the b coefficient indicates whether the relationship between the two variables is direct (positive) or inverse (negative). The regression is calculated such that the sum of squares of the difference between observed values of Y and the values of Y predicted by the regression equation (for a given value of X) are a minimum. The scatter of points around the regression line may be quantified using a dimensionless statistic, the coefficient of correlation (r). Values of r vary between $+1$ and -1. A value of 0 indicates that there is no correlation between X and Y. As the value of r approaches 1 ($+$ or $-$), the strength or statistical significance of the relationship increases. Table 2.1 provides critical values of r, for a given sample size (number of pairs of data), which allows us to accept or reject a given value of r as being statistically significant or not. The square of the correlation coefficient (r^2) tells us how good is the level of explanation achieved and is normally expressed as a percentage. The standard error (S_y) predicts that if a large sample of paired values of X and Y were obtained, about 68 per cent of these would fall within one standard error either side of the regression line, 95 per cent within two standard errors, and 99 per cent within three standard errors. Note that the title 'linear' regression model requires a linear relation between X and Y. If it is clear from a scattergram that the relationship is curvilinear, then a 'transformation' of one or both variables is required. This is most often achieved using the logarithm rather than the actual value of each observation. Further details of data transformations and of regression and correlation techniques in general are given in most texts which describe the use of statistical methods in geography. Some of these are listed under 'Further reading' at the end of this chapter. A high correlation shows that the relationship between X and Y is strong and sensitive; the regression model may therefore be used for prediction. Given values of X, Y can be predicted with confidence. Clearly the success

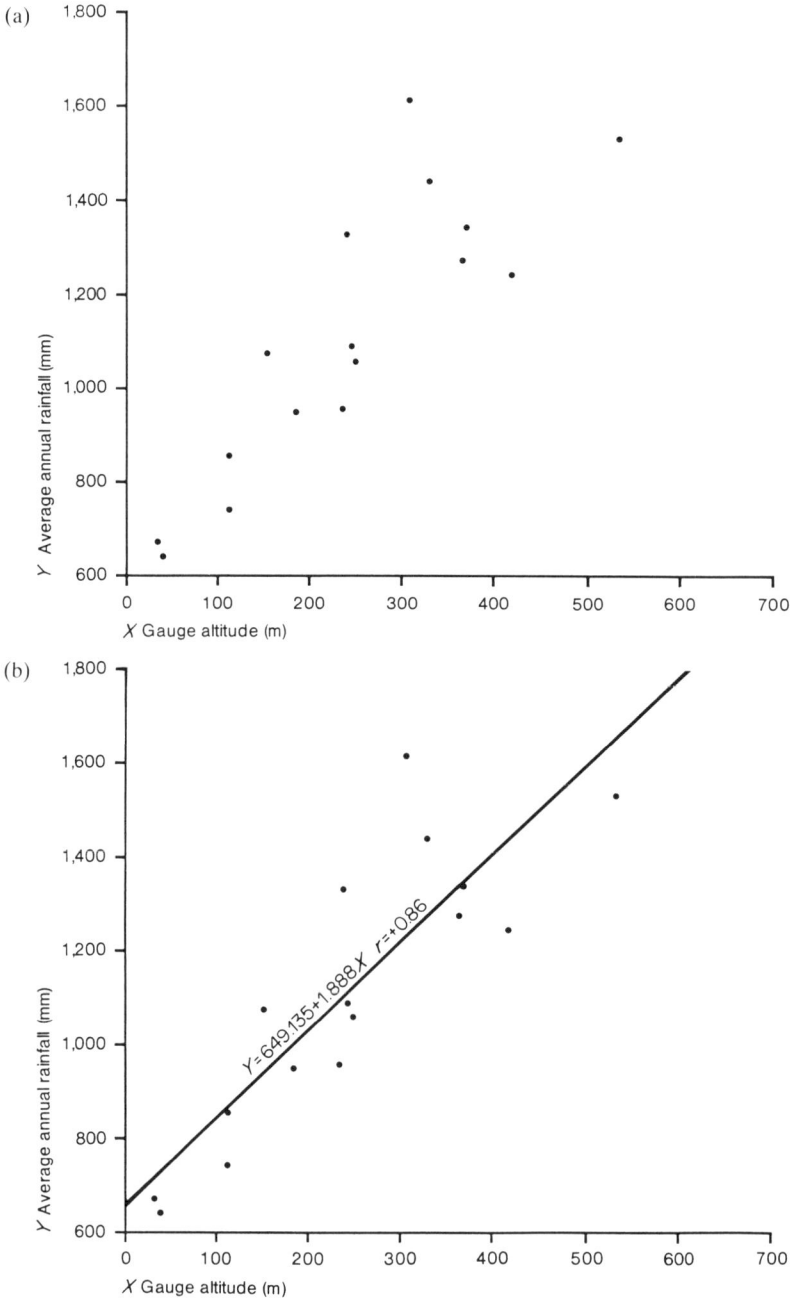

Figure 2.1 The relationship between raingauge altitude and average annual rainfall in the Southern Pennine hills, UK (a) scatter diagram; (b) best-fit regression line for the equation $Y = 649.135 + 1.888X$

Table 2.1   Significance levels for the correlation coefficient
r

Sample size (*n*)	Significance levels (*P*)		
	0.05	0.01	0.001
3	0.954	0.986	0.997
4	0.891	0.956	0.987
5	0.829	0.919	0.970
6	0.774	0.880	0.948
7	0.727	0.843	0.924
8	0.685	0.808	0.899
9	0.650	0.776	0.875
10	0.619	0.746	0.851
11	0.592	0.719	0.828
12	0.567	0.698	0.806
13	0.546	0.672	0.786
14	0.526	0.652	0.767
15	0.509	0.633	0.749
16	0.493	0.615	0.732
17	0.478	0.599	0.715
18	0.465	0.584	0.700
19	0.453	0.570	0.686
20	0.441	0.557	0.672
25	0.395	0.502	0.613
30	0.360	0.461	0.567
40	0.312	0.402	0.499
60	0.254	0.330	0.414
120	0.179	0.234	0.297

The correlation is significant if the calculated value of *r* is greater than the critical tabulated value at the chosen significance level (*P*) and appropriate sample size (*n*). Critical values of *r* were calculated using information contained in F. E. Croxton (1953) *Elementary Statistics*, Dover Press, New York, p. 312 and appendix V.

of the prediction falls as the correlation decreases towards zero. When values of *X* are used which lie above or below the range of the initial data set, the prediction is also less reliable since in this case the nature of the system may be radically different. Hidden thresholds may exist and different or additional processes may occur so that the relationship between *X* and *Y* may look very different to that originally defined.

Figure 2.2 provides a simple program (REG) to calculate the linear regression and correlation coefficient for a sample set of paired data. Table 2.2 give the set of data from which Figure 2.1 was drawn; this can be replicated if we select 'gauge altitude' as the independent variable, and

```
  10 REM BIVARIATE REGRESSION
  20 GOSUB 10000:REM INITIALIZE MACHINE
  30 GOSUB 18000
  40 A$="LINEAR REGRESSION AND CORRELATION": GOSUB 16000
  50 A$="*********************************":C=0:R=4
  60 C$="VARIABLE NAME X: ": A$="X": GOSUB 11000:
     GOSUB 19000:X$-A$
  70 C$="VARIABLE NAME Y: ":A$="Y": GOSUB 11000: GOSUB 19000:
     Y$=A$
  80 R=R+2:C$="Input number of pairs of data: ":A=3:
     GOSUB 12000:N=A
  85 XA=0:XB=0:YA=0:YB=0:Z=0
  90 DIM X(N),Y(N) :NX=1E9:XX=-NX:NY=NX:XY=XX
 100 FOR I=1 TO N
 101   PRINT"INPUT PAIR ";I:INPUT X,Y: X(I)=X: Y(I)=Y
 102   IF XX<X THEN XX=X
 104   IF NX>X THEN NX=X
 106   IF XY<Y THEN XY=Y
 108   IF NY>Y THEN NY=Y
 110   XA=XA+X:YA=YA+Y: XB=XB+X*X: YB=YB+Y*Y: Z=Z+X*Y
 120   NEXT I
 130 D=Z-XA*YA/N:E=XB-XA*XA/N:F=YB-YA*YA/N
 140 R=INT(D/SQR(E*F)*1000)/1000
 150 RA=INT(R*R*1000)/1000
 160 B=D/E
 170 XC=XA/N: YC=YA/N
 180 AA=YC-B*XC
 190 YD=SQR(YB/N-YC*YC)
 200 S=INT(YD*SQR(1-RA)*1000)/1000
 210 GOSUB 18000:PRINT"REGRESSION EQUATION IS:"
 215 A$="("+Y$+") ="+STR$(INT(AA*1000)/1000): IF B>0 THEN
     A$=A$+"+"
 220 A$=A$+STR$(INT(B*1000)/1000)+" x ("+X$+")": PRINT A$
 230 PRINT"CORRELATION COEFFICIENT = ";R
 240 PRINT"PERCENT EXPLANATION = ";RA
 250 PRINT"STANDARD ERROR (SY) = ";S
 290 C=0: R=8
 300 C$="How many predicted Y-values ?": A=0: GOSUB 12000:
     GOSUB 19000
 305 IF A=0 THEN GOTO 340
 310 NA=A: FOR I=1 TO NA
 315   C=0:C$= "Input X-value: ": GOSUB 12000
 320   Y=INT((AA+B*A)*1000)/1000
 330   C=20:A$="Predicted Y="+STR$(Y): GOSUB 16000:
       GOSUB 19000
 335   NEXT I
 340 C=0: GOSUB 19000:A$=" YOU WANT A GRAPH ?":GOSUB 16000:
     GOSUB 19000
 345 C$="IF YES PRESS <1>: IF NO PRESS <2>: ":A=1:
     GOSUB 12000:HH=A
 350 IF HH<>1 AND HH<>2 THEN GOTO 345
 355 IF HH=2 THEN 6999
 380 IF NX>0 THEN NX=0
 390 IF NY>0 THEN NY=0
 400 GOSUB 20000:REM SETS UP GRAPHICS SCREEN
 440 GOSUB 21000: GOSUB 22000
 450 FOR I=1 TO N:X=X(I):Y=Y(I):GOSUB 25000:NEXT I
 460 FG=0: FOR I=1 TO N
 470   X=X(I): Y=AA+B*X: GOSUB 26000: NEXT I
6999 PRINT"END OF RUN":END
```

Figure 2.2 A program to calculate linear regression and correlation (REG)

Table 2.2 Rainfall and altitude data for selected raingauge sites in the southern Pennine hills, UK

Site	Average annual rainfall (mm)	Gauge altitude (m)	Maximum altitude within 2 km (m)	Easting grid reference
Wessenden	1273	366	518	4068
Blackmoorfoot	1094	244	328	4090
Huddersfield	956	235	290	4113
Yateholme	1616	308	547	4111
Harden Moss	1345	369	475	4098
Wakefield	670	35	76	4327
Langsett	1059	250	370	4211
Underbank	949	184	355	4247
Cannon Hall	738	113	210	4273
Barnsley	640	40	118	4378
Chew	1536	532	541	4038
Bottoms Reservoir	1074	153	385	4023
Yeoman Hey	1329	239	503	4022
Dunford	1442	329	484	4151
Broomhead Moor	1246	418	490	4224
Moor Hall	856	124	340	4289

'average annual rainfall' as the dependent variable. The following output should result:

$$Y = 649.135 + 1.888X$$

$$r = 0.86 \qquad r^2 = 0.739 \qquad S_Y = 149.325 \qquad n = 16$$

The regression equation shows a strong direct relationship between altitude and rainfall. Average annual rainfall is estimated to be 649 mm at mean sea level ($X = 0$) and to increase by 188 m for every 100 m gain in height. Using the regression to make predictions of average annual rainfall, if we input the height of the Huddersfield gauge (235 m), we get a predicted value of 1092 mm, more than the recorded value but within one standard error. If we take Kinder Scout (the highest ground in the area with an altitude of 636 m), we predict an average annual rainfall of 1849 m. Since there are no records for this area, we cannot verify this estimate, but we note that the linear rainfall–altitude gradient is assumed to continue well beyond the range of our observations, a point which must be treated with a little caution unless other evidence is available to support this assumption.

There is much evidence to suggest that orographic rainfall is related more strongly to the generalized or 'smoothed' altitude of a range of hills, rather than to the specific heights of individual gauges. After all, if a gauge is in a deep valley, its rainfall is most likely to be determined by the height of the surrounding hills. We may be able to improve our correlation between altitude and average annual rainfall by using a different measure of height. Table 2.2 also includes a variable called 'height within 2 km'; this is the maximum height within a given radius around the gauge and, given the plateau form of the southern Pennines, is a good reflection of the average ground level near each raingauge. If this variable is used to correlate with average annual rainfall, it will be found that the correlation improves to $r = +0.94$, confirming the value of a generalized measure of altitude. The regression equation has a somewhat different form, however, and predicts that the rainfall at sea level is only 385 mm. Since we know that the east coast of England receives about 600 mm on average, this warns us of the difficulty of extrapolating our predictive equation beyond the range of observations, and in this case, to areas away from the study region. For the Huddersfield site our prediction is now 946 mm, very close to the observed value, and for Kinder Scout the predicted average if 1614 mm, which is more similar than the previous estimate to available records for sites on the high Pennine plateau such as Chew (Table 2.2).

The data given in Table 2.2 will allow us to build up a simple grey box analysis of Pennine rainfall. Since the hills have a western escarpment and dip towards the east, the easting grid reference correlates with rainfall totals $(r = -0.75)$, but as expected, altitude is a stronger control of rainfall. Since both measures of altitude are highly correlated, and because altitude decreases towards the eastern side of the study area, these three variables are highly interconnected, although strictly speaking, only 'height within 2 km' (as a surrogate for mean altitude around the gauge) is an independent variable. Figure 2.3 illustrates the correlation linkages for our very simple

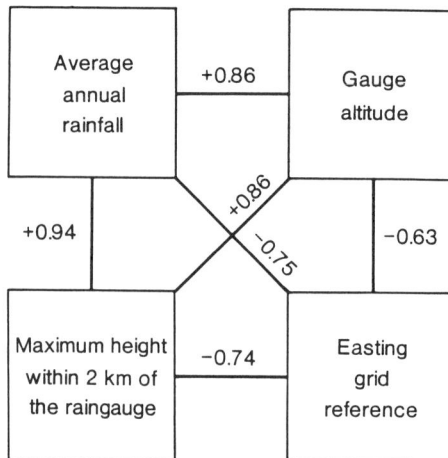

Figure 2.3 Constellation diagram for the Southern Pennine rainfall model

grey box model. Such diagrams are often termed 'constellation diagrams' and show well the morphology of the system. Of course, there is no information on the processes producing orographic rainfall, nor on the way in which different distributions of storm rainfall contribute to the overall picture. However, correlations and constellation diagrams are a useful starting point and often lead directly to more sophisticated types of model.

Of course, most environmental systems are much more complex than the simple bivariate relationships presented here. Many variables will be interrelated so that it becomes hard to evaluate which are the true independent variables, and which of those are the most important. To help us resolve these difficulties, the multiple regression test may be used. It is beyond the scope of this book to describe the use of this test, but many statistical texts contain a full account of the test and package programs are easily available for use on microcomputers. Some opportunity to use a multiple regression equation is given in the Unit Hydrograph program in Section 2.3. A further example of such a multivariate regression model is provided by the universal soil loss equation (USLE), in which soil erosion is computed on the basis of independent variables such as rainfall erosivity, slope length, soil erodibility, and so on. The USLE is described in detail in Mitchell and Bubenzer (1980) and would provide a neat problem for someone wishing to practise writing a simple computer model.

In some cases, use may be made of simple formulae which resemble regression equations, but which are not developed using formal statistical procedures. We may term such equations 'index models' and they provide one further type of morphological system. The parameters in these equations may be quantified subjectively, or, as in the following example, a method of weighting can be used. Just as with regression equations, index models are sometimes developed via nomograms, rather than with an equation; examples of these include the ASHRAE nomogram for estimating 'effective room temperatures' and some of the early flood prediction nomograms which were produced for certain rivers in the USA.

Figure 2.4 lists a program (INDEX) which uses the equation proposed by Davis (1968) to calculate an index of summer weather using temperature, sunshine and rainfall data. Davis was by no means the first climatologist to suggest such an index, but his parameters were produced on statistical grounds, rather than subjectively. This ensures that equal weight is given to each of the three variables involved. Davis's equation for an index of summer weather (I) is

$$I = 18T + 0.216S - 0.276R + 320$$

where T is the mean maximum daily temperature for June, July and August, in degrees centigrade,
            S is the total sunshine hours for the same three months, and
            R is the total rainfall in millimetres for these months.
It is clear that temperature and sunshine contribute positively to the index,

```
10 GOSUB 10000:REM INITIALIZES MACHINE:GOSUB 18000
20 DIM T(2), S(2), P(2),Q$(2),N$(2),K$(7)
30 RESTORE: FOR K=0 TO 7:READ K$(K):NEXT K:READ T$,Y$
40 FOR I=0 TO 2: READ Q$(I),N$(I),T(I),S(I),P(I):NEXT I
50 R=1:C=7:B$="*************************"
60 A$=B$:GOSUB 16000:R=2:GOSUB 16000
70 R=3:A$="*   SUMMER WEATHER INDEX   *":GOSUB 16000
80 A$=B$:R=4:GOSUB 16000:R=5: GOSUB 16000:C=0
90 FOR R=7 TO 14:A$=K$(R-7):GOSUB 16000:NEXT R  :R=23:
   GOSUB 19010:GOSUB 18000
100 C=0:R=0:C$="Station: ":A$=T$:GOSUB 11000:T$=A$
110 R=23:A$="                    ":GOSUB 16000
120 R=1:C$="   Year: ":A$=Y$:GOSUB 11000:Y$=A$  :R=3
130 A$="MONTH>>            JUNE  JULY  AUGUST":
    GOSUB 16000:R=4
140 A$=B$+LEFT$(B$,13):GOSUB 16000:R=5:F=0
150 C=0:IF F=0 THEN FOR R=5 TO 7:A$=Q$(R-5):GOSUB 16000:
    NEXT R
160 FOR I=0 TO 2
170   R=5:C=20+6*I:C$="":A$=STR$(T(I)):GOSUB 11000:
    T(I)=VAL(A$)
180   R=6:A$=STR$(S(I)):GOSUB11000:S(I)=VAL(A$)
190   R=7:A$=STR$(P(I)):GOSUB11000:P(I)=VAL(A$)
200   NEXT I:F=1
210 C=0:R=10:C$="<A> to accept: <C> to change: ":A$="":
    GOSUB 11000
220 B$=A$:C=29:A$="   ":GOSUB 16000
230 IF B$="C" THEN R=5:GOTO 150
240 IF B$<>"A" THEN GOTO 210
250 TT=0:SS=0:PP=0
260 FOR I=0 TO 2:TT=TT+T(I):SS=SS+S(I):PP=PP+P(I): NEXT I
270 I=18*(TT/3)+(20/92)*SS-0.276*PP+320
280 C=0:R=12:A$="Davis' Summer Weather Index": GOSUB 16000
290 C=20:R=14:A$="  = "+STR$(INT(I+.5))+"  ":GOSUB 16000
300 R=23:C=0:A$="Modify data from the top":GOSUB 16000
310 GOTO 100
320 END
330 REM ========================
340 DATA This index model calculates a,summer weather index
    using the
350 DATA Davis (1968) equation.,The calculations require :
360 DATA "     Mean max. daily temp. (Centigrade)",
    "     Total sunshine (hours)"
370 DATA "     Total rainfall (mm)","for the months of June,
    July and August"
380 DATA OXFORD,1987
390 REM Data for OXFORD 1987:Headings,Month,Temp,Sun,Rf
400 DATA Mean Temp (deg C),June,20,300,12
410 DATA Total hrs Sunshine,July,17,145,35
420 DATA Total rainfall (mm),August,15,230,75
430 REM +++++++++++++++++++++++++++
```

Figure 2.4   A program to calculate summer weather indices (INDEX)

whilst rainfall reduces its value. The higher the summer weather index, the better the summer has been!

Some sample data is listed in Tables 2.3 and 2.4 for selected stations in the United Kingdom for the summers of 1976 (the 'drought' year) and 1977 (an 'average' summer). These data were extracted from the *Weather Log* which is published by the journal *Weather* and which can be used as a

Table 2.3 Summer weather data for selected stations in the United Kingdom for 1976

Station	Month	Mean max. temp (°C)	Total sunshine	Total rainfall
Plymouth	June	20.2	255	11
	July	21.8	259	27
	August	23.3	308	15
Kew	June	24.8	268	8
	July	26.0	287	25
	August	24.5	269	13
Lowestoft	June	20.6	314	9
	July	21.6	309	19
	August	20.7	273	57
Birmingham	June	22.5	260	21
	July	24.0	254	26
	August	23.1	223	62
Valley	June	18.2	216	16
	July	20.7	225	30
	August	22.4	291	11
Durham	June	20.3	213	25
	July	22.1	237	30
	August	21.9	221	18
Dyce	June	18.7	158	19
	July	21.1	239	15
	August	20.3	216	35
Stornaway	June	14.5	116	101
	July	17.3	145	88
	August	17.1	251	16
Eskdalemuir	June	17.4	111	68
	July	20.8	213	85
	August	20.3	217	25
Oxford	June	24.4	263	17
	July	25.9	256	14
	August	24.4	258	24

continuing source of data. If you have access to records from your local meteorological observatory, these can be used to compare different summers (1985 seems to have been an awful summer in England – but how does it compare with previous ones using the Davis index?). Using the tabulated data, the summer weather indices given in Table 2.5 result. As expected, 1976 has much more favourable indices, particularly in the south of England where the drought was most severe. In the north the effect of the drought was much less marked, and at Stornaway (Outer Hebrides) the indices are almost the same. It is interesting too that Eskdalemuir, in the Southern

Table 2.4 Summer weather data for selected stations in the United Kingdom for 1977

Station	Month	Mean max. temp (°C)	Total sunshine	Total rainfall
Plymouth	June	15.7	163	55
	July	19.8	223	42
	August	19.2	211	106
Kew	June	16.8	141	55
	July	21.5	223	11
	August	19.7	138	127
Lowestoft	June	15.6	114	22
	July	18.8	145	13
	August	18.3	158	64
Birmingham	June	15.7	153	118
	July	19.8	191	9
	August	18.5	140	66
Valley	June	16.0	233	74
	July	19.3	222	21
	August	19.2	211	55
Durham	June	15.9	159	63
	July	18.8	136	9
	August	18.6	141	38
Dyce	June	14.7	173	57
	July	18.0	166	58
	August	17.2	170	101
Stornaway	June	13.1	195	57
	July	15.3	150	51
	August	16.1	190	57
Eskdalemuir	June	16.0	210	95
	July	18.5	197	49
	August	17.5	184	115
Oxford	June	16.7	125	87
	July	21.3	194	6
	August	19.8	123	129

Uplands, had a less impressive summer in 1976 than Durham, which is within 50 miles, but much lower and on the east coast. The index of 938 for Kew in 1976 confirms the extreme nature of that summer in south-east England.

As can be seen from Figure 2.4, models such as INDEX are relatively simple to program, and yet can provide some useful simulation exercises. We can use 'real' data to compare stations, or to compare different years at the same station. We can also use 'synthetic' data to investigate the importance of each parameter to the overall prediction. This latter process is known as sensitivity analysis and is discussed in some detail in section 7.3, as is the experimental use of simulation models in general.

Table 2.5 Summer weather indices for selected climate stations in the United Kingdom for 1976 and 1977

Station	1976	1977
Plymouth	875	721
Kew	938	723
Lowestoft	868	699
Birmingham	867	695
Valley	831	750
Durham	831	704
Dyce	794	670
Stornaway	668	658
Eskdalemuir	739	688
Oxford	921	701

2.3 SCHEMATIC MODELS

It is often possible to portray a complex system using simple linear equations. Whilst this may represent a gross simplification of reality, the exercise is still valuable in that it forces us to try to model the theoretical basis of the system. Using our simple model we can then investigate the role of individual variables within the system in a controlled manner. In the last section we were concerned with essentially bivariate relationships. However, in complex environmental systems many more variables must be considered: since we cannot easily set up lots of controlled field experiments to find out how a single factor influences a process, a simulation model helps us to investigate such controls theoretically. Once we have an idea of how the system is structured, we may then progress to more complex models, as described in Chapters 3 and 4. As will be seen in the examples below, even at this stage our models must be physically realistic, but because the system is greatly simplified, with process mechanics not correctly modelled, they remain black box models.

2.3.1 Flood hydrographs using the time–area method

Consider the way in which a flood hydrograph is produced at the outlet of a drainage basin. The basin has two components: a certain proportion of the rainfall input is converted to run off on the hillslopes; this runoff then enters the stream and is routed through the channel network to the basin outlet.

Let us first model the conversion of rainfall into runoff. In 1850, an Irish land drainage expert called Mulvaney first suggested an extremely simple method of predicting runoff, known as the rational method:

$$Q = ciA$$

where Q is the discharge, i is the rainfall intensity, A is the catchment area, and c is a runoff coefficient. c has a range between 0 and 1, representing the fraction of rainfall which is converted into runoff; c may be as low as 0.01 in very permeable basins, and as high as 0.9 in impermeable urban catchments. Clearly the formula is extremely simple: no reference is made to actual runoff processes occurring within the catchment, and the runoff coefficient is considered constant through time. Figure 2.5(a) shows a drainage basin divided into ten zones on the basis of isochrones – a line of equal travel time along the channel, usually set at one hour intervals. Thus the runoff produced in area A_1 takes one hour to reach the basin outlet; runoff from area A_2 takes two hours to reach the outlet, and so on. Thus, if we have a series of hourly rainfall totals (i_1, i_2, i_3 . . . i_n), the total discharge for any given hour will be

$$Q_t = c_1 A_1 i_{(t-1)} + c_2 A_2 i_{(t-2)} + c_n A_n i_{(t-n)}$$

Figure 2.6 lists a program (TIMAREA) which allows you to produce a hydrograph for a basin divided into a maximum of 20 time zones. The area and runoff coefficient may be selected for each area. Initially you need not worry about the absolute value of the numbers being used: treat the simulation as purely schematic (a later paragraph will describe how to apply the model to real catchments). Sample input data is provided in Table 2.6 with the resulting hydrograph shown in Figure 2.5(b). Though each subunit of the basin is a very simple system, when a number of subunits are combined to yield the complete drainage basin, a complex system results. In particular, the shape and runoff response of the basin may be varied in a controlled manner. Try these two simulations:

1   For a given basin shape and rainfall input sequence, change the value of c at different places in the basin, and examine its effect on the hydrograph. If you greatly increase c, this may simulate the effect of deforestation or of urbanization, for example.
2   For a fixed rainfall sequence, fixed number of areas with fixed values of c, keep the total basin area constant, but vary the distribution of the area. This causes basin shape to vary: examine the effect of changing basin shape on the hydrograph.

Of course, you can alter the total basin area, and also change the rainfall inputs. This could give some very haphazard results if you vary the parameters at random, so we suggest the use of controlled simulation experiments to ensure that you can determine cause and effect. The time–area method is a very simple method of hydrograph prediction and is seldom used today by engineers for flood prediction in large rural catchments, although the method is still applied in urban flood analysis. If you wish to apply this formula to a real basin, some idea of the length of each time zone (measured along the

(a)

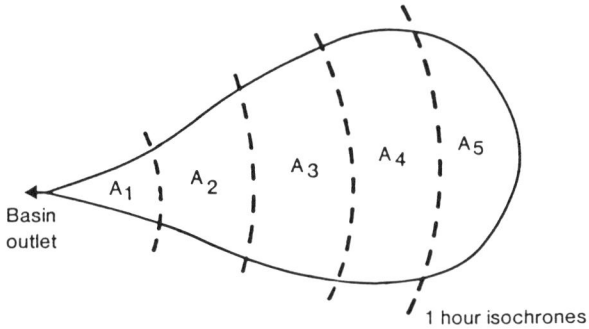

(b)

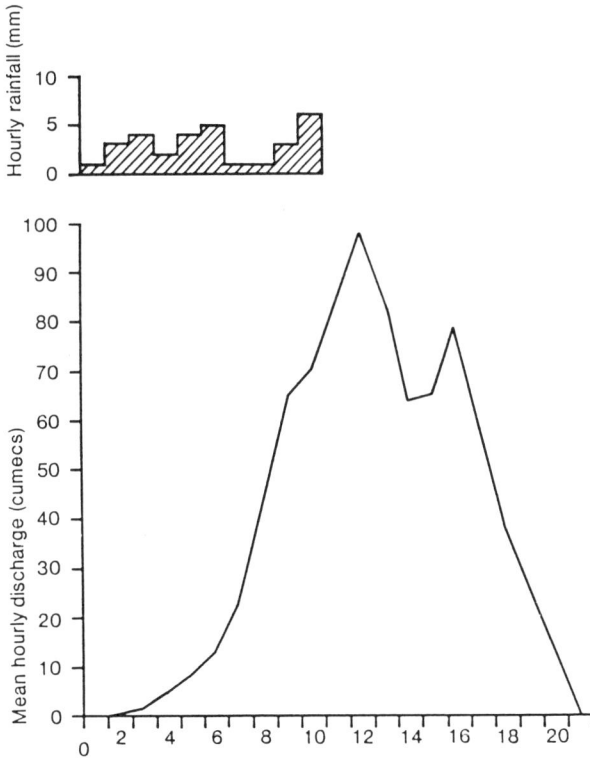

Figure 2.5 The time–area method of hydrograph prediction (a) drainage basin divisions; (b) sample rainfall input and hydrograph output for data provided in Table 2.6

```
 10 GOSUB 10000:REM Initialize
 20 RESTORE:B$="*":FOR I=1 TO 5:B$=B$+B$:NEXT I:PRINT B$:
    PRINT B$
 30 PRINT "TIME AREA HYDROGRAPH SIMULATION":PRINT B$:
    PRINT B$:C=0:R=6
 40 C$="Enter No of sub-unit areas: ":GOSUB 510
 50 Z=X:R=R+1:DIM A(Z),C(Z)
 60 A$="SUB-AREAS (1 nearest outflow)":GOSUB 16000:R=R+1
 70 FOR I=1 TO Z
 80   C=0:A$=STR$(I)+":   ":GOSUB   16000:C=5:IF   I=1:
    GOSUB 510:GOTO 100
 90   GOSUB 540
100   A(I)=X:NEXT I
110 R=R+5:GOSUB 550:R=R-4:IF R<0 THEN R=0
120 C=0:A$="SUB-AREA RUNOFF COEFFICIENTS":GOSUB 16000
130 R=R+1:FOR I=1 TO Z
140   C=0:A$="A("+STR$(I)+")  of "+STR$(A(I))+"sq km: ":
    GOSUB 16000:C=20:IF I=1 THEN GOSUB 510:GOTO 160
150   GOSUB 540
160   C(I)=X:NEXT I
170 GOSUB 18000:R=0:C=0:GOSUB 510
130 NR=X:NQ=Z+X:DIM R(NR),Q(NQ)
190 R=2:XR=0:FOR I=1 TO NR
200   C=0:A$="Hour "+STR$(I)+": ":GOSUB 16000:C=12
210   IF I=1 THEN GOSUB 510:GOTO 240
220   GOSUB 540
230   IF XR<X THEN XR=X
240   R(I)=X:NEXT I
250 REM =================
260 REM Main calculation
270 XQ=0:GOSUB 18000:PRINT B$:
    PRINT"          OUTFLOW HYDROGRAPH":PRINT B$
280 XQ=0:FOR J=1 TO NQ+1
290   S=0:FOR I=1 TO Z
300     IF J-I>0 AND J-I<=NR THEN S=S+R(J-I)*A(I)*C(I)
310     NEXT I
320   S=S/3.6:Q(J-1)=S:IF S>XQ THEN XQ=S:REM Convert to l/s
330   NEXT J:C=0:R=4:A$="Maximum discharge =":GOSUB 16000
340 C=21:B=4:D=2:A=XQ:GOSUB 17200
350 R=6:GOSUB 580:REM Output
360 REM End main section
370 REM =================
380 REM Graphical output
390 GOSUB 20000:PRINT "RUNOFF";
400 NX=0:XX=NQ:NY=0:XY=XQ*1.5:GOSUB 22000:REM Set up graph
    grid
410 X=NQ*.7:Y=XY/20:A$="Time (Hrs)":GOSUB 23000
420 X=XX/3:Y=XQ*.4:A$="Discharge (l/s)":GOSUB 23000
430 FG=0:FOR X=0 TO NQ
440   Y=Q(X):GOSUB 26000:NEXT X
450 PRINT" AND RAINFALL"
460 X=XX/10:Y=XY*.9:A$="Rainfall (mm)":GOSUB 23000
470 KR=XY/3/XR:FOR X=0 TO NR-1
480   XA=XA+1:YA=XY:Y=YA-KR*R(X+1)
490   GOSUB 27000:NEXT X
500 GOSUB 29500:END
510 READ A,LL,UL
520 READ C$:IF C$="END" OR C$="End" OR C$="end" THEN RETURN
530 C$="Enter "+C$+" ("+STR$(LL)+"-"+STR$(UL)+"): "
540 GOSUB 12000:IF A<LL OR A>UL THEN GOTO 540
550 X=A:R=R+1:IF R<23 THEN RETURN
560 R=24:A$="":GOSUB 16000:PRINT:R=22
```

(*continued*)

```
570 GOSUB 18000:R=0:RETURN
580 FOR J=1 TO NQ
590   RF=0:IF J<=NR THEN RF=R(J)
600   C=0:A$="HR "+STR$(J):GOSUB 16000
610   C=6:A$=": RF(mm) "+STR$(RF):GOSUB 16000
620   C=17:A$=": DISCH(1/s)":GOSUB 16000
630   C=33::A=Q(J):GOSUB 17200
640   R=R+1:IF R=23 THEN GOSUB 680
650   NEXT J:GOSUB 680:RETURN
660 REM ===============
670 REM Prints or waits before clearing screen
680 GOSUB 29000:IF F7=2 THEN C=0:R=23:
    C$="Press RETURN to continue ":A$="":GOSUB 11000
690 GOSUB 18000:R=0:RETURN
700 REM ===============
710 DATA 5,1,20,No of sub-unit areas
720 DATA 10,0,100,Sub-area
730 DATA 0.3,0,1,C
740 DATA 8,2,50,No of hours rainfall
750 DATA 5,0,100,Rf
```

Figure 2.6 A program to calculate flood hydrographs using a time–area model (TIMAREA)

Table 2.6 Input data for the time–area hydrograph prediction model (the resultant hydrograph is depicted in Figure 2.3(b))

Hour	Rainfall (mm)	Area number	Size (km^2)	c
1	1	1	10	0.2
2	3	2	15	0.2
3	4	3	20	0.2
4	2	4	20	0.2
5	4	5	25	0.2
6	5	6	30	0.8
7	1	7	30	0.8
8	1	8	20	0.8
9	3	9	15	0.8
10	6	10	10	0.8

stream channel) may be obtained from the formula suggested by Kirpich (1940):

$$L = \frac{T_c S^{1.25}}{0.00025}$$

where T_c is the time of concentration (usually one hour), S is the channel gradient and L is the length of each time–area subunit. You will need to allocate drainage areas to each length of channel, and a value of c to each area, and then you can proceed to produce some tentative flood hydrographs for your river.

2.3.2 The unit hydrograph model

This method of flood prediction was suggested by Sherman (1932). It is a graphical method of predicting the outflow hydrograph for a basin and requires no knowledge of the runoff processes operating. The flood hydrograph observed at the basin outlet is separated into quickflow (i.e. storm runoff) and baseflow. The volume of quickflow is divided by the basin area to yield the depth of excess rainfall over the basin which produced the flood. The unit hydrograph is calculated by dividing the ordinate of the quickflow hydrograph by the excess rainfall so that the hydrograph now represents the runoff which would result from a unit amount of excess rainfall. Traditionally, in imperial units, the one-inch unit hydrograph was used, but nowadays 25 mm or 10 mm units are used. Having done this, the unit hydrograph is simply multiplied by the rainfall excess to yield the flood prediction. The unit hydrograph is usually calculated as an average of many individual unit graphs, each of which resulted from a single input of rainfall in a given period of time. Once produced, the unit hydrograph may be used to predict the flood response for complex rainfall events: successive rainfall inputs (for successive unit time periods) are first converted into excess rainfall by estimating the infiltration loss. Then each excess rainfall amount is multiplied by the unit graph. The resultant hydrographs are then added together, each successive one being delayed by the unit time interval.

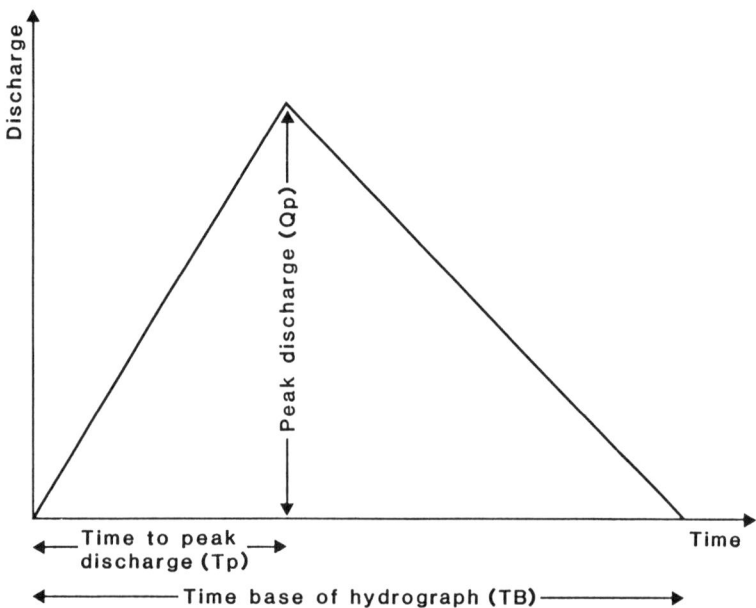

Figure 2.7 The dimensions of a synthetic unit hydrograph

Where no records of river discharge exist, we cannot produce a unit hydrograph by the normal graphical method. Snyder (1938) calculated 'synthetic' unit hydrographs: here the unit graph is simplified to a triangle whose dimensions depend on its time base (TB), the time to peak discharge (T_p) and on the magnitude of the peak discharge (Q_p) itself (Figure 2.7). Snyder's method has been updated for England and Wales by the *Flood Studies Report* (NERC, 1975) who calculate the unit hydrograph dimensions on the basis of four catchment characteristics – main channel length (L); channel gradient (S); regional flood index (RSMD); and the fraction of the basin which is urban (URBT). Figure 2.8 lists a program (UNIT) which allows you to calculate the unit hydrograph for any basin using these four basin properties. Table 2.7 gives appropriate data for a number of basins of differing size and characteristics throughout England and Wales. This also demonstrates the use of a multiple regression equation (in the calculation of T_p) and you can use the procedure to investigate how each basin characteristic affects the unit hydrograph properties. For example, would a longer main channel produce a larger or smaller unit peak discharge? Would a larger urban area increase the dimensions of the unit hydrograph? Try answering these questions, plus others of your own, using the first part of the UNIT program. The relevant equations are as follows (from *Flood Studies Report*, NERC, 1975):

$$T_p \text{ (hours)} = 46.6 L^{0.14} S^{-0.38} (\text{URBT})^{-1.99} (\text{RSMD})^{-0.4}$$

$$Q_p \text{ (cumecs/100 km}^2) = \frac{220}{T_p}$$

$$\text{TB (hours)} = 2.52 T_p$$

If you wish to simulate a real outflow hydrograph for any basin, you must input additional information: a soil index (which ranges from 0 for very permeable soils to 1 for an impermeable catchment); each hour's rainfall total for your 'design' storm; an antecedent soil wetness index (which ranges from a saturated value of 0 to a dry value of 125); and daily rainfall totals for the five days prior to the storm. Using a chosen design storm (e.g. an extreme storm for which you have obtained the hourly rainfall totals), you can simulate the response of different basins to the rainfall input. For one basin, you can investigate the effect of varying basin characteristics (such as the fraction of urban area) on the resultant hydrograph.

The unit hydrograph is a classic example of a black box model. It contains no process information and yet has proved quite satisfactory for flood prediction over the last half century. The model does, however, have severe limitations: it assumes a linear relation between rainfall and runoff, a constant time base for flood runoff, and is unable to accommodate major changes in basin properties such as deforestation or land drainage. Thus, whilst the unit hydrograph model can provide quite acceptable flood predictions for some

```
 10 GOSUB 10000:REM INITIALIZE MACHINE
 20 GOSUB 18000
 30 REM 10MM/1 HOUR UNIT HYDROGRAPH
 40 REM BASED ON THE FLOOD STUDIES REPORT (1975)
 50 PRINT:PRINT TAB(8);"UNIT HYDROGRAPH MODEL"
 60 PRINT"*************************************":PRINT
 70 C=0: R=4: C$="INPUT CATCHMENT NAME: ": A$="":
    GOSUB 11000: W$=A$: GOSUB 19000
 80 C$="AREA (Sq Km): ":LL=0: UL=500: A=100:
    GOSUB 12500:AA=A:GOSUB 19000
 90 C$="LENGTH (KM): ":A=25:GOSUB 12500: L=A: GOSUB 19000
100 C$="SLOPE (%): ":UL=100: A=5:GOSUB 12500: S=A:
    GOSUB 19000
110 C$="5yr-1day Rf (RSMD)":UL=300:A=150: GOSUB 12500:
    RE=A: GOSUB 19000
120 C$="FRACTION URBAN":UL=1: A=0:GOSUB 12500:U=A+1:
    GOSUB 19000
130 REM calculate unit hydrograph dimensions
140 LET TP=46.6*(L^0.14)*(S^-0.38)*(U^-1.99)*(RE^-0.4)
150 QP=((220/TP)*(AA/100)):TB=2.52*TP
160 GOSUB 18000: PRINT"10MM/1 HOUR UNIT HYDROGRAPH":
    PRINT"FOR";W$;" CATCHMENT "
170 PRINT:PRINT"TIME TO PEAK = ";TP;" HOURS"
180 PRINT"PEAK DISCHARGE = ";QP;" CUMECS"
190 PRINT"TIME BASE = ";TB;" HOURS"
200
210 R=7: C=0: A$="Enter Initial for choice": GOSUB 16000:
    R=R+1
220 C=5: A$="<F> for a Flood Hydrograph": GOSUB 16000
230 R=R+1: A$="<U> for another Unit Hydrograph": GOSUB 16000
240 C=5: R=10: C$="<Q> to Quit: ": A$="F": GOSUB 11000
250 IF A$="Q" OR A$="q" THEN GOTO 960
260 IF A$="U" OR A$="u"THEN GOSUB 18000: GOTO 70
270 IF A$<>"F" AND A$<>"f" THEN GOTO 240
280 REM CALCULATE U.H. HOURLY ORDINATES
290 R=R+2: C=0: C$="SOIL INDEX":LL=0: UL=1:A=.1:GOSUB 12500:
    SL=A: GOSUB 19000
300 DIM Q(100):TA=INT(TP):IF TA=0 THEN GOTO 320
310 FOR I=1 TO TA:Q(I)=(QP/TP)*I:NEXT I
320 TC=INT(TB):FOR I=(TA+1) TO TC:
    Q(I)=(QP/(TB-TP))*(TB-I):NEXT I
330 REM CALCULATE STANDARD PERCENT RUNOFF(SPR):
    SPR=95.5*SL+0.12*U
340 C$="SOIL MOISTURE DEFICIT":UL=125:A=50:GOSUB 12500:
    SMD=A: GOSUB 19000
350 GOSUB 18000
360 DIM RF(100):XQ=0
370 C=0: R=0: A$="ENTER HOURLY Rf's (<=100 mm)":
    GOSUB 16000
380 R=1: A$="End sequence with a minus value":
    GOSUB 16000: R=3
390 H=0:P=0:A=0:XR=0
400 C$="HOUR "+STR$(H+1)+" Rainfall (mm): ":GOSUB 12000
410 IF A>100 THEN GOTO 400
420 GOSUB 19000: IF A<0 THEN GOTO 450
430 H=H+1: RF(H)=A: P=P+A:IF A>XR THEN XR=A
440 GOTO 400
450 GOSUB 19000: DIM AD(5)
460 A$="PREVIOUS 5 DAYS RAINFALL TOTALS(MM) ?":
    GOSUB 16000
470 GOSUB 19000:FOR J=1 TO 5
480   UL=100:C$="Day "+CHR$(48+J)+": Rainfall (mm)":A=0
490   GOSUB 12500: AD(J)=A:GOSUB 19000
500   NEXT J
```

(*continued*)

```
510 API5=(0.5^0.5)*(AD(1)+(0.5^AD(2))+(0.5^2*AD(3))
    +(0.5^3*AD(4))+(0.5^4*AD(5)))
520 REM CALCULATE CATCHMENT WETNESS INDEX
530 CWI=125+API5-SMD
540 SPR=95.5*SL+1.2*U
550 LET PR=SPR+0.22*(CWI-125)+0.1*(P-10)
560 REM CONVOLUTE TUH WITH NET RAINFALL PATTERN
570 GOSUB 18000:PRINT"OUTFLOW HYDROGRAPH"
580 C=0:R=2:A$="TIME":GOSUB 16000
590 C=13:A$="RAINFALL":GOSUB 16000
600 C=26:A$="DISCHARGE":GOSUB16000
610 DIM QQ(TC*2+H),RO(TC*2+H)
620 FOR J=1 TO TC:QQ(J)=0:NEXT J
630 FOR J=(TC+1) TO (TC+TC):QQ(J)=Q(J-TC):NEXT J
640 FOR J=(TC*2+1) TO (TC*2+H):QQ(J)=0:NEXT J
650 R=3:FOR J=(TC+1) TO (TC*2+H)
660   FOR K=1 TO H
670     IF J>=K THEN N=N+((RF(K)*PR/1000)*QQ(J-K+1))
680     NEXT K:RO(J)=N:N=0
690   L=J-TC:RO(J)=INT(RO(J)*1000)/1000
700   IF XQ<RO(J) THEN XQ=RO(J)
710   GOSUB 990: NEXT J
720 PR=INT(PR*1000)/1000
730 CWI=INT(CWI*1000)/1000
740 CWI=INT(CWI*1000)/1000
750 GOSUB 19000:A$="RUNOFF PERCENT = "+STR$(PR):
    GOSUB 16000: GOSUB 19000
760 A$="CATCHMENT WETNESS INDEX = "+STR$(CWI): GOSUB 16000:
    GOSUB 19000
770 GOSUB 19000: A$="Should graph include Unit Hydrograph ?":
    GOSUB 16000: GOSUB 19000
780 C$="<Y> YES: <N> NO: then <RETURN>: ": A$="":
    GOSUB 11000: GOSUB 19000
790 IF A$="Y" OR A$="y" THEN A=1:GOTO 820
800 A=2:IF A$<>"N" AND A$<>"n" THEN GOTO 780
810 REM ==================
820 GOSUB 20000:REM SETS UP GRAPHICS SCREEN
830 A$="Runoff forecast for "+W$:GOSUB 24000
840 NX=0:XX=(TC+H)*1.2
850 NY=0:XY=XQ*1.1:GOSUB 21000
860 GOSUB 22000: FG=0
870 A$="Disch": X=XX/50:Y=XY: GOSUB 23000
880 A$="Hours": X=XX*.6: Y=XY/10: GOSUB 23000
890 FOR J=(TC+1) TO (TC*2+H)
900   K=J-TC:X=K:Y=RO(J):GOSUB 26000:NEXT J
910 Y=Y+XY/20:A$="Runoff":GOSUB 23000
920 IF A=2 THEN GOTO 950
930 FG=0:X=0:Y=0:GOSUB 26000:X=TP:Y=XQ:GOSUB 26000:
    X=TB:Y=0:GOSUB 26000
940 X=TP:Y=XQ-XY/20:A$="Unit Hydrograph x "
    +STR$(INT(XQ/QP*100+.5)/100):GOSUB 23000
950 FOR XA=1 TO H:X=XA-1:YA=XY:Y=XY*(1-RF(XA)/XR/2):
    GOSUB 27000:NEXT XA:A$="Rain":GOSUB 23000
960 PRINT"END OF RUN":END
970 REM ==================
980 REM Printout routine
990 B=3:C=3:A=L:GOSUB 17300
1000 D=1:C=16:A=RF(L):GOSUB 17200
1010 B=4:D=3:C=30:A=RO(L):GOSUB 17200
1020 GOTO 19000
1030 REM ++++++++++++++++++
```

Figure 2.8   A program to compute synthetic unit hydrographs (UNIT)

Table 2.7 Selected basin characteristics for use in the unit hydrograph model (data abstracted from the *Flood Studies Report* (NERC, 1975), with permission)

Gauging station	FSR number	Area (km²)	L	S	RSMD	URBT	SOIL	AAR
Rhymney	57001	63	17.2	14.7	55.2	0.050	0.490	1530
Chew	53004	130	22.9	3.6	35.5	0.000	0.355	1013
Vyrnwy	54003	94	12.2	29.2	77.8	0.000	0.465	1908
Piddle	44002	183	31.6	3.3	41.5	0.000	0.177	973
Exe (S)	45002	422	48.1	5.7	50.5	0.000	0.326	1402
Exe (T)	45001	601	68.0	4.4	45.6	0.003	0.293	1280
Dart	46003	248	35.2	6.5	71.9	0.000	0.361	1821
Roding	37001	303	62.6	1.2	15.9	0.081	0.409	635
Calder	27029	342	31.1	5.1	48.5	0.059	0.425	1219
Severn	54022	8	4.8	63.5	81.8	0.000	0.500	2449
Windrush	39006	365	60.4	1.9	25.7	0.003	0.209	785
Findhorn	7001	417	47.3	9.3	57.1	0.000	0.500	1337
Frome	53007	262	27.7	2.3	32.8	0.010	0.330	983

Exe stations: (S) = Stoodleigh; (T) = Thorverton

rivers, more versatile models must be process based, especially if they are to be applied to ungauged basins or are to be used to assess the impact of land use changes. Such models will be described in later chapters.

2.3.3 Modelling solutional denudation on hillslopes

It is not just in hydrology that simulation models can prove valuable. A further example of a schematic model is provided from hillslope geomorphology. Carson and Kirkby (1972) suggest a simple model which describes the erosion of a hillslope by solutional processes. The basis of the model is shown in Figure 2.5. Unit-area runoff at any point on the slope is determined by the difference between rainfall and evaporation. Since soil moisture content increases downslope, so does evaporation; thus unit-area runoff decreases downslope (Figure 2.9(a)). Initial solutional loss is proportional to the product of rainfall (rf) and the fraction of the oxide present in the soil (p). If soluble oxides are present (i.e. p is close to one), then evaporation almost certainly will mean that the soil-water solution becomes saturated with solute, and excess solute is precipitated. In this case the solutional loss if proportional to the unit-area runoff (ro). Thus for any point on the slope, solutional loss is proportional to (rf.p) or to (ro) whichever is less (Figure 2.9(b)). For limestones ($p > 0.9$), loss is proportional to runoff; thus solutional erosion decreases downslope and slope decline occurs. For igneous rocks ($p < 0.3$) in humid areas (i.e. high runoff to rainfall ratio), solutional loss is proportional to (rf.p) and parallel retreat occurs. As the runoff to rainfall ratio falls, the resulting pattern of erosion on igneous rocks provides

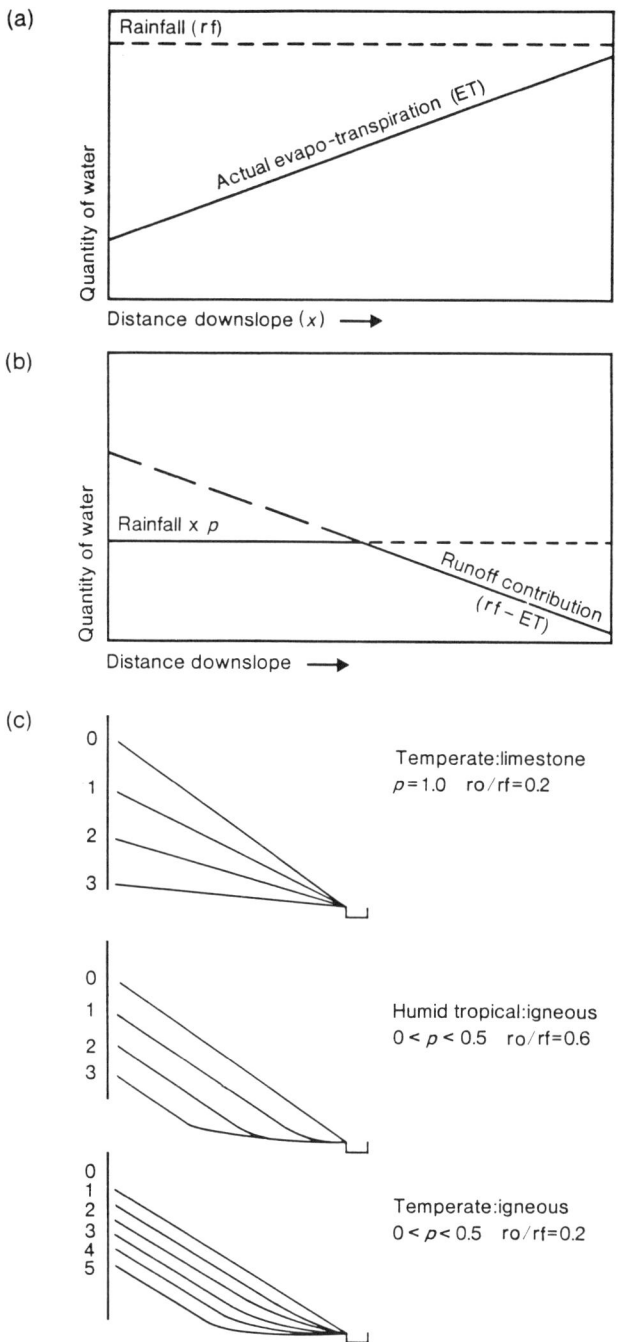

Figure 2.9 A conceptual model for solutional denudation of hillslopes (a) rainfall, evaporation and unit runoff; (b) patterns of solutional denudation in relation to rainfall and runoff; (c) generalized patterns of hillslope evolution for selected rock types and climatic regimes (based on Carson and Kirkby, 1972)

38

parallel retreat at the top of the slope, with slope decline lower down (Figure 2.9(c)).

Figure 2.10 lists a program (CARSON) which reproduces this model of solutional erosion of hillslopes. No reference is made to the thermodynamics of solutional processes: the approach is purely schematic. Nevertheless, the model simulates realistically the evolution of an initially straight slope section. Controlled simulation experiments will allow you to study the role of individual variables in slope development, in this case: the fraction of oxide present in the rock; the overall evaporation rate (this controls the magnitude of the unit area runoff); and, the rate at which evaporation increases downslope. The initial tabulation of erosion rates shows the pattern of erosion over the slope, and the decrease in slope height is printed, and later plotted, for a series of 100 iterations of the model.

2.4 ANALYSIS OF OBSERVED TRENDS

For any input/output system, the role of 'time' is crucial. We may distinguish between the ephemeral compensating changes associated with a system which is in dynamic equilibrium and the irreversible changes involved in system evolution. It is clearly important to be able to analyse input and output data to look for evidence of 'timeless reversible' fluctuations in system behaviour and to distinguish these from the 'timebound irreversible' changes associated with evolution (terminology from Chorley and Kennedy, 1971, Chapter 7). From such black box analysis may come the identification of particular trends or periodicities which may point the way towards more detailed study of processes and storages within the system.

Figure 2.11 lists a program (RUN1) which analyses a time series of observations to identify three types of trend (these three types of temporal variation are shown schematically in Figure 2.12):

1   Running means are used to eliminate small-scale oscillations so that longer-term trends or periodic cycles show up. The most common method of smoothing is to use a running mean. This removes short-term fluctuations and emphasizes longer-term trends. For example, for monthly temperature data, the annual cycle can be removed by use of a 12-month running mean. There are several possible disadvantages of running means including loss of marginal data and creation of false or reversed peaks and troughs. However, if interpreted with care, this method of smoothing time series data can identify long-term trends and cycles, whether regular or irregular. When using your own data, RUN1 allows you to select the period over which the running mean is calculated; there is no advantage in choosing too long a period since this reduces the number of mean values calculated. For the sample data sets, the monthly data uses a 12-month average, and the annual values use an 11-year mean. The original data set and the running mean are tabulated and then plotted automatically.

```
 10 GOSUB 10000:REM INITIALIZE MACHINE
 20 K$="***": FOR K=1 TO3: K$=K$+K$:NEXT K: S$="          ":
    K$=S$+K$
 30 PRINT K$: PRINT K$
 40 PRINT S$;"HILLSLOPE SOLUTION MODEL"
 50 PRINT K$: PRINT K$
 60 DIM DIS(10),RO(10),H(10),M(10,10)
 70 GOSUB 520: REM Initial Slope profile
 80 R=8: C=0: C$="SOLUBLE ROCK FRACTION":LL=0:UL=1:A=0.4:
    GOSUB 12500:P=A
 90 RAIN=2
100 GOSUB 19000: C$="SUMMIT EVAP (mm)": UL=RAIN::A=RAIN/2:
    GOSUB 12500: EVAP=A
110 GOSUB 19000: A$="Evap increases downslope in arid areas":
    GOSUB 16000
120 GOSUB 19000: C$="BASAL EVAP (mm)":LL=EVAP:UL=RAIN*2-EVAP:
    A=RAIN:GOSUB 12500: GG=A
130 SUM=RAIN-EVAP: FOR J=1 TO 10
140    DIS(J)=RAIN*P
150    RO(J)=RAIN-EVAP-(GG-EVAP)*J/10:SUM=SUM+RO(J):
       IF SUM<0 THEN RO(J)=RO(J)-SUM: SUM=0
160    IF DIS(J)>RO(J) THEN DIS(J)=RO(J)
170    NEXT J
180 GOSUB 18000:R=0
190 A$="Pos'n":GOSUB 16000
200 FOR J=1 TO 10:C=3+J*3:B=3: A=J:GOSUB 17300: NEXT J:R=R+1
210 C=0:A$="Eros'n": GOSUB 16000
220 FOR J=1 TO 10:C=3+J*3:B=3:A=DIS(J)*50:GOSUB 17300:
    NEXT J: R=R+2
230 C=0: A$="TIME          H E I G H T S _____":
    GOSUB 16000: R=R+1
240 FOR W=0 TO 100
250    C=0: A$=STR$(W)+":":GOSUB 16000
260    IF INT(W/10)*10<>W THEN GOTO 290
270    FOR J=1 TO 10
280      M(W/10,J)=H(J):C=3+J*3:B=3:A=H(J): GOSUB 17300:
         NEXT J:R=R+1
290    IF H(1)=H(10) THEN GOTO 360
300    FOR J=1 TO 10
310      H(J)=H(J)-DIS(J)/3: IF H(J)<0 THEN H(J)=0
320    NEXT J
330    FOR J=1 TO 9
340      IF H(J)<H(J+1) THEN H(J+1)=H(J)
350    NEXT J
360    NEXT W
370 GOSUB 19010
380 GOSUB 20000:REM SETS UP GRAPHICS SCREEN
390 FG=0:REM INITIALIZES GRAPH
400 NX=0:XX=10
410 NY=0:XY=M(0,1):GOSUB 21000
420 SX=2:GOSUB 22000
430 X=XX/50:Y=XY:A$="Elev": GOSUB 23000
440 X=XX*.8:Y=XY/10: A$="Dist.":GOSUB 23000
450 FOR K=0 TO 10
460    GOSUB 18000: PRINT "TIME ";K*10;
470    FG=0:FOR J=1 TO 10:X=J:Y=M(K,J):GOSUB 26000:NEXT J
480    NEXT K
490 END
500 REM ==========================
510 REM Initial profile Elev  subroutine
520 FOR J=1 TO 10: H(J)=70-5*J:NEXT J: RETURN
530 REM ==========================
```

Figure 2.10 A program to model solutional denudation on hillslopes (CARSON)

```
>LIST
  10   REM TIME SERIES
  20   GOSUB 10000:REM INITIALIZE MACHINE
  30   DIM A(200),Z(200)
  40   PRINT"TIME SERIES ANALYSIS"
  50   PRINT"*******************":C=0: R=4
  60   R=4:A$="<1> to enter your own data": GOSUB 16000
  70   R=5: C$="<2> to use data on file: ":A=2: GOSUB 12000
  80   IF A=2 THEN GOTO 190
  90   IF A<>1 THEN GOTO 70
 100   GOSUB 18000: R=0: C$="Enter File Name: ":A$="":
       GOSUB 11000
 110   IF LEN(A$)>7 THEN GOTO 100
 120   F$=A$: GOSUB 14500
 130   C$="Enter Number of values: ": A=0:R=2: GOSUB 12000:
       IF A>200 THEN GOTO 130
 140   GOSUB 14800
 150   N=A: A=0:FOR I=1 TO N
 160     C$="Value "+STR$(I)+": ":GOSUB 12000: GOSUB 19000
 170     GOSUB 14800: NEXT I: GOSUB 14700
 180   GOSUB 18000: GOTO 60
 190   GOSUB 18000: R=0: C=0
 200   GOSUB 14550
 210   R=24: C$="Enter file name: ":A$="": GOSUB 11000: F$=A$
 220   GOSUB 18000: R=0:GOSUB 14600
 230   GOSUB 14950:   GOSUB 16000: R=2
 240   GOSUB 14900: N=A: A$=STR$(N)+" Observations":
       GOSUB 16000: R=4
 250   FOR I=1 TO N
 260     GOSUB 14900: Z(I)=A: A=N-I:GOSUB 17000
 270     NEXT I: GOSUB 14700
 280   A$="Enter No of periods for running means":
       GOSUB 19000:GOSUB 16000
 290   C=0:GOSUB   19000:  C$="Number":LL=1: UL=24: A=12:
       GOSUB 12500: P=A
 300   GOSUB 19000: A$="Working on calculations at 0":
       GOSUB 16000
 310   M=INT(P/2): C=27
 320   FOR I=1 TO (N-P)
 330     A=I: B=3: GOSUB 17300
 340     FOR J=I+1 TO (I+P-1)
 350       T=T+Z(J):NEXT J
 360     T=T+0.5*Z(I)+0.5*Z(I+P)
 370     A(I+M)=T/P:T=0:NEXT I
 380   GOSUB 18000:PRINT F$:PRINT"Running mean for ";P;" time
       periods"
 390   C=0:R=4:A$="TIME":GOSUB 16000
 400   C=10:R=4:A$="OBSERVED":GOSUB 16000
 410   C=20:R=4:A$="RUNNING MEAN":GOSUB 16000
 420   PRINT:PRINT:R=6
 430   FOR J=(M+1) TO (N-P+M)
 440     C=0:B=3:A=J:GOSUB 17300
 450     C=10:B=3:D=2:A=Z(J):GOSUB 17200
 460     C=20:B=3:D=2:A=A(J):GOSUB 17200
 470     PRINT:GOSUB 19000:NEXT J
 480   NQ=Z(1):XQ=Z(1)
 490   FOR I=1 TO N
 500     IF NQ>Z(I) THEN NQ=Z(I)
 510     IF XQ<Z(I) THEN XQ=Z(I)
 520     NEXT I
 530   GOSUB 19010
 540   GOSUB 20000:REM SETS UP GRAPHICS SCREEN
```

(*continued*)

```
550    FG=0:REM INITIALISES GRAPH
560    NX=0:XX=N
570    NY=NQ:XY=XQ
580    GOSUB 21000
590    GOSUB 22000
600    FOR J=1 TO N:X=J:Y=Z(J):GOSUB26000:NEXT J
610    FOR J=(M+1) TO (N-P+M):X=J:Y=A(J):GOSUB25000:NEXT J
620    GOSUB 18000: R=0:C=0: A$="Enter <L>  for a linear
       trend plot": GOSUB 16000
630    R=1: C$="<C> for a cosine curve: <Q> to quit: ":
       A$="L":GOSUB 11000
640    IF A$="Q" OR A$="q" THEN GOTO 940
650    IF A$="C" OR A$="c" THEN GOTO 870
660    IF A$<>"L" AND A$<>"l" THEN GOTO 630
670    REM BIVARIATE REGRESSION
680    GOSUB 18000: PRINT "LINEAR REGRESSION"
690    XA=0:XB=0: YA=0: YB=0: Z=0:FOR I=1 TO N
700       XA=XA+I:YA=YA+Z(I):XB=XB+I*I:YB=YB+Z(I)*Z(I)
710       Z=Z+I*Z(I):NEXT I
720    D=Z-((XA*YA)/N):E=XB-(XA*XA)/N:F=YB-(YA*YA)/N
730    R=D/SQR(E*F)
740    RA=R*R
750    B=(INT((D/E)*100000)))/100000
760    XC=XA/N:YC=YA/N
770    AA=YC-B*XC
780    YD=SQR(YB/N-YC*YC)
790    S=YD*SQR(1-RA)
800    A$="Y = "+STR$(INT(AA*1000+.5)/1000)
810    IF B>0 THEN A$=A$+"+"
820    A$=A$+STR$(INT(B*1000+.5)/1000)+" * X": PRINT A$
830    A$="CORRELATION COEFFICIENT = "+STR$(INT(R*1000+.5)/1000):
       PRINT A$;
840    FG=0:X=0:Y=AA+B*X: GOSUB 26000
850    X=N:Y=AA+B*X:GOSUB 26000
860    GOTO 940
870    REM ROUTINE TO FIT A COSINE CURVE
880    FOR I=1 TO N:TT=TT+Z(I):XB=XB+Z(I)*Z(I):NEXT I
890    GOSUB 18000: R=0: A$="FITTING COSINE CURVE":
       GOSUB 16000: R=1
900    C$="PHASE SHIFT":LL=0: UL=1: A=0:GOSUB 12500: F=X
910    MY=TT/N:SD=SQR(XB/N-MY*MY):FG=0
920    FOR I=1 TO N:Y=MY+SD*COS((((I/P)*2*PI)-((F*12/P)*2*PI)):
       X=I:GOSUB 26000:NEXT I
940    PRINT"End of Run":END
```

Figure 2.11   A program to analyse time series (RUN1)

2   Linear trends in the time series can be identified using regression
     analysis. RUN1 offers the option of plotting the best-fit regression line
     for the original data set and the regression equation and correlation
     coefficient are printed. Some caution is needed in the interpretation of
     any linear trend since the trend may not be statistically significant, or
     the actual trend may be irregular rather than linear and may mask
     periodic variations over the time period. For the nitrate data set, a linear
     trend may have real meaning since stream concentrations are increasing
     in response to higher fertilizer inputs.

42

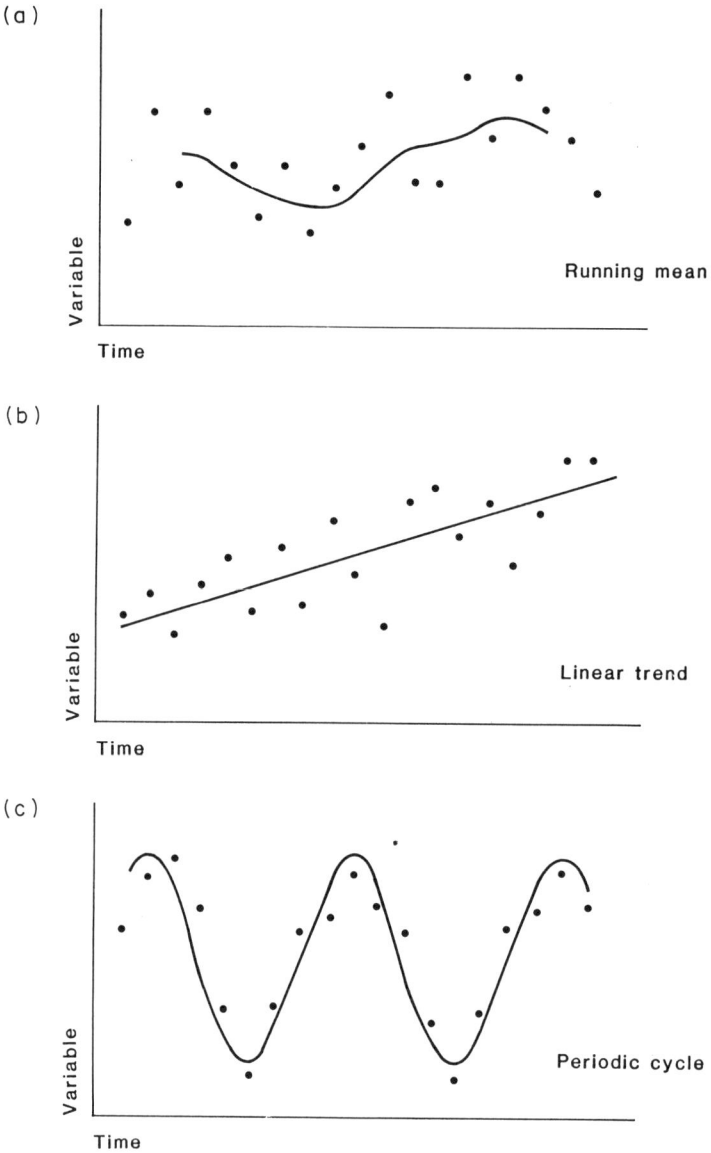

Figure 2.12 The three types of temporal variation identified by program RUN1

3   Regular periodic variations, such as diurnal or annual cycles, may be
     identified using some form of harmonic analysis. RUN1 plots a cosine
     curve with the same wavelength as the period length used for calculating
     the running mean. No regression equation is provided but visual
     inspection will show if cycles are evident. The general form of the cosine
     curve is

$$Y = M + SD\cos(2\pi - k)$$

where M is the mean of the input data, SD the standard deviation, and k is the 'phase shift' – this defines how long after the beginning of each cycle the peak value in the cosine curve will occur. Note that you cannot plot both the linear trend and the cosine curve on the same program run.

Three data sets are provided as files to use with RUN1 for you to experiment with before using your own data. These are:

1 Monthly rainfall totals for Slapton (South Devon) from January 1974 to December 1984 (in millimetres): file name SLAP_RF.
2 Monthly values of the mean concentration of nitrate-nitrogen in the Slapton Wood stream from January 1971 to October 1985 (in milligrams per litre): file name SLAP_NI.
3 Mean annual air temperatures for the Radcliffe Meteorological Observatory at Oxford from 1815 to 1984 (in degrees centigrade): file name OX_TEMP.

You may analyse those original data sets, or provide data of your own, which can be written to another file. The suggested period of 12 for running means is appropriate to remove cyclic elements from monthly data, but can be changed as you think fit for each particular data set.

Using the data provided, or your own observations, use program RUN1 to identify temporal variations. Look for evidence of cycles, regular or irregular, and for the presence of trends. Look also for major deviations from the general pattern, such as droughts or times of abnormally high solute leaching. Having identified these properties of the time series, we need to consider the causes of these changes and the effects that might be produced. Such questions lead us to consider the processes operating and demand an approach that goes beyond black box analysis. The value of time series analysis is that it provides a starting point: given the complexity of environmental systems, we need to know what changes are occurring before we can start to understand them.

2.5 CONCLUSIONS

Black box models provide a much simplified view of reality. Usually the models are empirically derived, particularly when regression techniques are employed. Where models are theoretically based, as with the solutional denudation model, the mathematical structure of the model is kept very simple since often the modeller has little idea of how the system operates in detail. However, by running a large number of simulations, the user can build up some idea of the way in which the system operates, and of the limits of the system. From these ideas, the user may go on to formulate more sophisticated but realistic models, examples of which appear in the next two chapters.

2.6 FURTHER READING

Carson, M. A. and Kirkby, M. J. (1972) *Hillslope Form and Process*, Cambridge University Press, Cambridge.

Chorley, R. J. and Kennedy, B. A. (1971) *Physical Geography: A Systems Approach*, Prentice Hall, London.

Davis, N. E. (1968) An optimum summer weather index, *Weather,* **23**, 305–17.

Kirpich, Z. P. (1940) Time of concentration of small agricultural watersheds, *Civil Engineering,* **10**(6), 362.

Mitchell, J. K. and Bubenzer, G. D. (1980) Soil loss estimation. In M. J. Kirkby and R. P. C. Morgan (eds.) *Soil Erosion*, Wiley, Chichester, 17–62.

Mulvaney, T. J. (1850) On the use of self-registering rain and flood gauges, *Transactions of the Institute of Civil Engineers of Ireland,* **4**(2), 1–8.

Natural Environment Research Council (1975) *Flood Studies Report*, 5 volumes, Institute of Hydrology, Wallingford, England.

Sherman, L. K. (1932) Streamflow from rainfall by the unit-graph method, *Engineering News Record,* **108**, 501–5.

Snyder, F. F. (1938) Synthetic unitgraphs, *Transactions of the American Geophysical Union,* **19**(1), 447–58.

CHAPTER 3

Process Models

3.1 INTRODUCTION

In the last chapter, it was shown how simple linear regression analysis could be used as a model to link the input to and output from a process. Thus, variables were seen to be related in either a simple linear and additive, or a log-linear and multiplicative, fashion, with the choice of variables being determined by an appreciation of the processes under consideration. In this chapter, this awareness is taken a step further to include the mechanisms underlying these processes and how they might be expressed mathematically. It will become clear that this understanding of processes helps to direct field work in the choice of variables and that the relationships uncovered are very different from those allowed for in the simple regression model.

3.2 PREDICTION OF RUNOFF

One of the examples used in the previous chapter was that of the unit hydrograph method for predicting drainage basin runoff. Here, the three regression equations for time to peak, percentage runoff and average baseflow were used to calculate hydrograph size and shape. In these equations, the variables describing the soil characteristics of the basin and the antecedent soil moisture conditions are simply used additively or multiplicatively to provide the parameters of the hydrograph. This clearly does not reflect the role of these variables in governing the runoff amounts and, not surprisingly, the predictive power of this method is often poor (NERC, 1975). This chapter provides an alternative model in which the hydrological processes at work are simulated, albeit in a rather simplified manner.

As shown in Figure 3.1, the components of the hydrological cycle (Ward, 1975) can be viewed as a series of storages and flows. This simple diagram can be used to form the conceptual framework for a numerical model to describe a particular part of the cycle. Take the soil moisture store (Figure 3.2) for example. This illustrates two very important concepts in modelling – first, the idea that the amount of water in storage (h in Figure 3.2) determines the rate of outflow (Kirkby, 1975). The analogy with a bucket

45

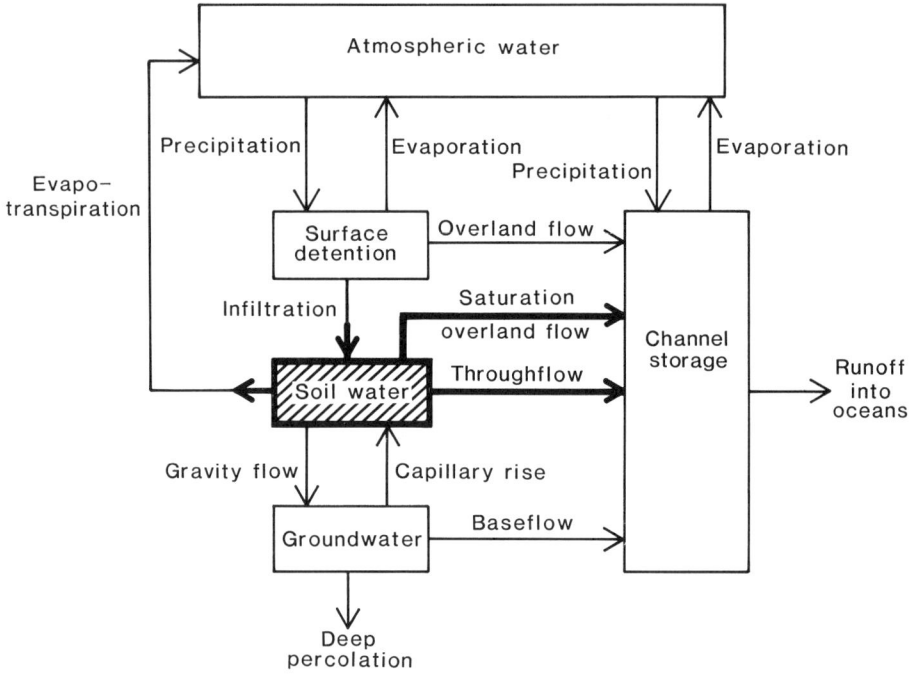

Figure 3.1   The hydrological cycle expressed as a series of water storages and flows; storage and flows shown emphasized are those modelled in STORFLO

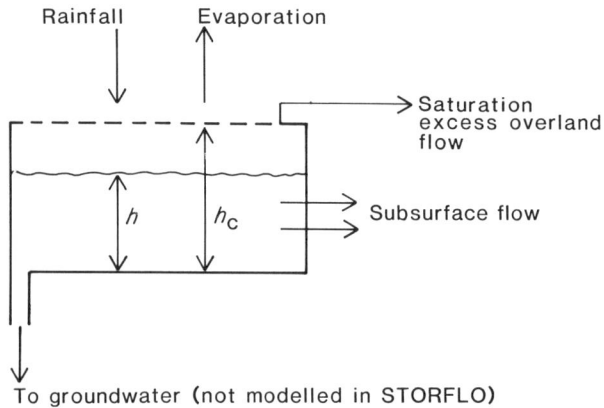

Figure 3.2   Conceptual soil moisture store; h is storage level in the soil, h_c is storage capacity of the soil

of water with a hole in the bottom is obvious – the more water in the bucket, the faster the outflow. The second idea is that of storage capacity (h_c in Figure 3.2). In other words, continuing the analogy with the bucket, there is a certain limit to the amount of water that can be poured into the bucket before it overflows. This is also one of the ways in which the soil behaves (Dunne and Black, 1970) and is especially important in humid temperate areas where rainfall rates are almost always less than the infiltration rate allowed by the soil.

This simple description provides some idea as to *how* the soil and soil moisture mediate between the rainfall input to the soil and the outflow to the river channel. These ideas now need to be couched in a mathematical form. Before doing this, however, it is necessary to think about the application of the model to the field and the scale at which it might reasonably be expected to work.

First, the model is limited to the type of catchment where the hydrology is dominated by simple throughflow and saturated overland flow responses – for example the classic catchment of East Twin in the Mendips (Weyman, 1974). The model as it stands cannot cope with Hortonian overlandflow or pipeflow. Secondly, the aim of the model is to generate streamflow and thus we might expect the soil moisture store to represent the entire soil cover of the catchment. Furthermore, the description above has not recognized any effect of time dependence. The model is, therefore, limited to small drainage basins where delivery times of water over land to the channel are small and where there is no need to consider the problem of flood-wave propagation within the channel before the outflow point is reached. To be consistent with this restriction, the time scale of the model is taken to be days and the output is in terms of daily flow. In turning the concepts discussed above into an operational model, the rate of drainage from the soil store can be represented by

$$Q_t = Q_0 \exp\left(\frac{S_t}{m}\right) \tag{3.1}$$

where Q_t is rate of outflow at time t (mm/day)
   Q_0 is rate of outflow at saturation (mm/day)
   S_t is soil moisture *deficit* (mm)
   m is a model parameter representing the soil characteristics

This is an exponential model for soil drainage and is represented by the series of curves in Figure 3.3 for different values of m. (For a full discussion of the derivation of this model see Kirkby, 1975.)

Because the soil moisture decreases as the water drains out of the soil, a second equation is required to account for the quantities of water involved. In other words, a water balance equation is also needed. (For a full discussion of mass balance models see Chapter 4.) This equation can be expressed as follows:

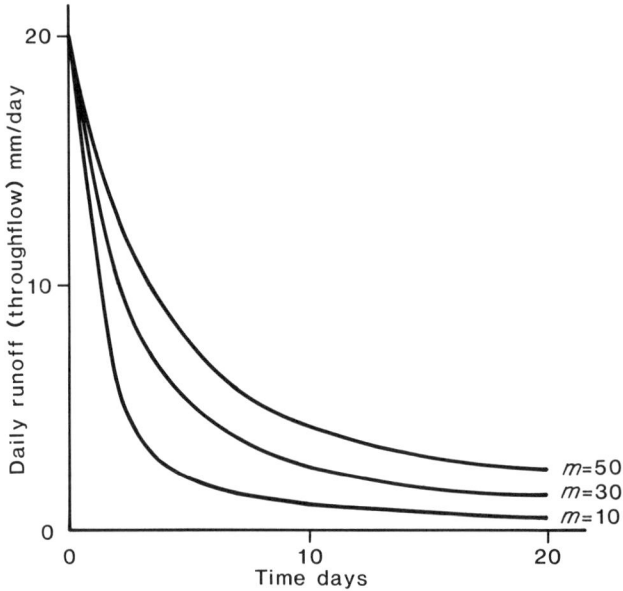

Figure 3.3   The effect on the recession curve of varying parameter m

$$\frac{\mathrm{d}S}{\mathrm{d}t} = R - E - Q \qquad (3.2)$$

where $\mathrm{d}S/\mathrm{d}t$ is change in soil moisture deficit over the day (mm/day)
   R is daily rainfall (mm/day)
   E is daily evaporation (mm/day)
   Q is total runoff during the day (mm/day)

If moisture input is greater than output, i.e. $R > (E+Q)$, then the moisture content of the drainage basin increases, the soil moisture deficit decreases and $\mathrm{d}S/\mathrm{d}t$ is positive.

Putting the equations (3.1) and (3.2) together and integrating (Kirkby, 1975) gives

$$Q_{(t+1)} = (R-E)/\{1-\exp\left[-\frac{(R-E)}{m}\right] + (R-E)\exp\left[-\frac{(R-E)}{m}\right]/Q_t\} \quad (3.3)$$

where Q_t is discharge rate at time t (mm/day)
   $Q_{(t+1)}$ is discharge rate at time $t+1$ (mm/day)

In addition to these basic equations which govern the drainage of the soil, the idea of storage capacity has also to be incorporated. The maximum value of Q_0 (in equation (3.1)) may be thought of as the saturated hydraulic conductivity of the soil. If the throughflow rate given by equation (3.3) exceeds this maximum rate then the excess must occur as overland flow and is, therefore, given by the equation

$$Q_{OF} = R - E - Q_0 - S \qquad (3.4)$$

where S is the moisture taken into storage (mm/day). Overland flow is assumed to take place very rapidly and, therefore, on the time scale of one day may be considered to run off instantaneously, i.e. once generated, overland flow is added directly to the total runoff for the time period.

Having discussed the conceptual and mathematical basis for the model, it is now possible to formulate the computational outline of the model. Figure 3.4 shows a flow diagram of the calculations involved.

A simple program, STORFLO, to perform these calculations is given in Figure 3.5. It is written so that it stores the values of rainfall and runoff to be tabulated and output graphically using the subroutines given in Appendix 1 for machine specific operations. The arrays r(20) and qd(20) are used for a 20-day period of rainfall and runoff respectively. Model parameters such as initial discharge, discharge at saturation and the recession parameter m are input at the beginning of the model run. As explained above, m controls the rate of drainage of the catchment and can be thought of in terms of variables such as soil texture, etc. One of the uses of the model might be to find out how streamflow varies with m and, work on linking m to measurable soil or drainage basin parameters might be a useful field project associated with the computing exercise.

Clearly, the model discussed above represents just one small part of the hydrological cycle (Figure 3.1) but, using it as an example, it has been possible to move quite a long way towards specifying a more realistic, process-based model compared to the unit hydrograph model given in Chapter 1. Just three simple concepts were needed – the relationship between storage and flow, the idea of storage capacity and a mass balance equation. The same type of approach can be applied to other parts of the cycle and, with modification, to other space and time scales. For guidance in this choice, you are referred to Chapter 8.

3.3 PREDICTION OF EVAPO-TRANSPIRATION

In the previous model, the loss of water by evaporation was assumed to be constant. This section now looks in detail at the *process* of evaporation as it is affected by the interaction of vegetation and atmosphere. This example illustrates another analogy used in modelling processes – that of electrical resistance.

Again, there are 'black box' approaches to the problem of predicting evapo-transpiration. One particular example is the Thornthwaite formula for potential evapo-transpiration (Dunne and Leopold, 1978):

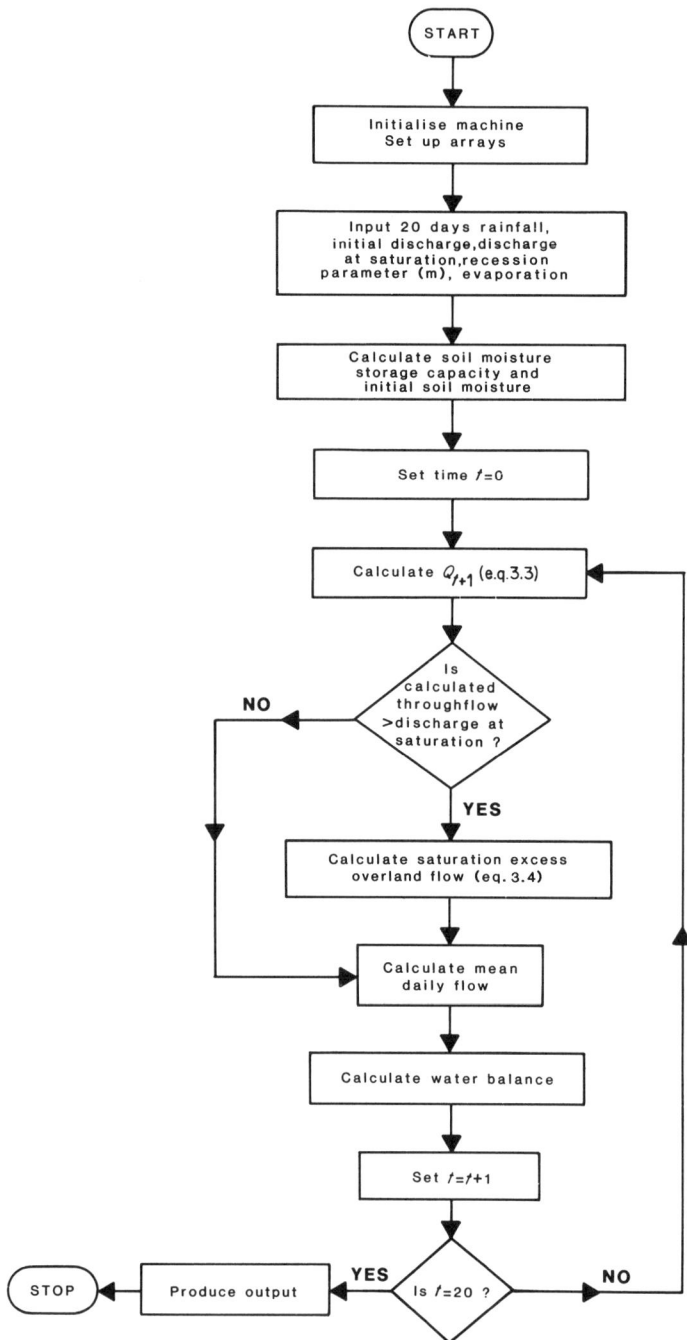

Figure 3.4 Flow diagram showing sequence of calculations used in predicting runoff from rainfall in STORFLO

$$E_t = 1.6 \left(\frac{10T_a}{I} \right)^a \tag{3.5}$$

where E_t is potential evapo-transpiration in cm/month

T_a is mean monthly air temperature (degrees C)

I is annual heat index $\left(\sum_{i=1}^{12} (T_{ai}/5)^{1.5} \right)$

$a = 0.49 + 0.01791I - 0.0000771I^2 + 0.000000675I^3$

An empirical equation such as this one has a number of drawbacks. First, the atmospheric effect is assumed to be very simple. Secondly, there is no explicit influence of vegetation – the values will be the same regardless of whether there is a cover of grass or of trees and this is clearly not the case (e.g. see Clarke and Newson, 1978). Thirdly, the value of evapo-transpiration given is that of the potential loss, whereas the actual loss will be dependent on the soil moisture. Fourthly, there is no consideration given to interception losses incurred. The model developed here addresses just the first two of these issues.

Work by Penman (1948) and Monteith (1973) has refined the analysis of evapo-transpiration in two ways – first, by understanding the heat transfers involved in evaporating moisture from leaf to air, and secondly, by considering the processes by which moisture moves from the plant on to the leaf surface via the stomata. Electrical resistance forms the best analogy for the processes involved. Resistance to the movement of water on to the plant surfaces is controlled by soil moisture, plant growth, etc. and is expressed by r_s – the bulk surface resistance. The calculation of r_s takes many different forms dependent on the type of crop as is demonstrated in the program given below (Meteorological Office, 1981). Resistance to the movement of water from the surface of the plant into the air is given by the aerodynamic resistance, r_a, which is dependent on turbulence in the atmosphere and is given by

$$r_a = \frac{\ln(2/z_0)^2}{k^2 u} \tag{3.6}$$

where z_0 is the roughness length of the crop $(0.1h)$

h is crop height (m)

k is Von Karman's constant (0.41)

u is wind speed some 2 m above the crop (m/s)

Finally, transpiration is calculated using the Penman–Montieth equation (Monteith, 1973) in which

$$E = \frac{D(R_n - G) + \rho c_p (e_s - e)/r_a}{L[D + g(1 + r_s/r_a)]} \tag{3.7}$$

where E is rate of water loss (kg/m²/s)

D is rate of change of saturated vapour pressure with temperature (mb/°C)

```
 10 GOSUB 10000:REM initialize machine
 20
 30 PRINT:PRINT TAB(8);"RAINFALL-RUNOFF MODEL":PRINT:REM title
 40
 50 REM set up arrays for rainfall and runoff (20 days max)
 60 REM r rain; qd runoff; of overlandflow; th throughflow
 70 DIM r(20),qd(20),of(20),th(20)
 80
 90 PRINT"Rainfall is input in line 1000":
    PRINT"Do you want to edit this (Y/N)? ":A$="N":
    A$=INKEY$(3000):IF A$="Y" THEN END
100 RESTORE:n=20:FOR i=1TOn:READ r(i):NEXT i:
    REM Reads rainfall data (20 days)
110
120 PRINT:PRINT"Input model parameters:"
130 PRINT:PRINT"Enter a number between 0 and 20":LL=0:UL=20:
    C=0:R=9:A=5:C$="Initial discharge (mm/day) = ":GOSUB 12510:
    q=A:i=q
140 PRINT:PRINT"Enter a number between 0 and 20":R=12:A=15:
    C$="Saturation discharge (mm/day) = ":GOSUB 12510:qo=A
150 PRINT:PRINT"Enter a number between 1 and 60":LL=1:UL=60:
    R=15:A=1:C$="Recession parameter m = ":GOSUB 12510:m=A
160 PRINT:PRINT"Enter a number between 0 and 5":LL=0:UL=5:R=18:
    A=1:C$="Evaporation (mm/day) = ":GOSUB 12510:e=A
170 PRINT:PRINT"Press RETURN to continue":INPUT a$:GOSUB 18000
180
190 REM set storage capacity and initial soil moisture
200 A=qo:GOSUB 14100:sm=m*A:A=i:GOSUB 14100:s=sm-m*A:
    REM nat.logs
210
220 REM calculate discharge at end of day
    (uses correction to avoid errors in exponentiation)
230 FOR t=1 TO n:qi=q:a=r(t)-e:IF a=0 THEN a=-0.1
240   b=EXP(-a/m):q=a/(1-b+a/qi*b):IF a=-0.1 THEN a=0
250
260   REM calculate saturation excess overland flow
270   A=qi:GOSUB 14100:IF q>qo THEN of(t)=a-(qo+qi)/2-(sm-m*A):
      q=qo
280
290   REM calculate mean daily flow
300   th(t)=(q+qi)/2:qd(t)=th(t)+of(t)
310
320   REM calculate water balance
330   qt=qt+qd(t):rt=rt+r(t):tt=tt+th(t):ot=ot+of(t)
340   NEXT t:REM next day
350
360 REM calculate final soil moisture and total evaporation
370 A=q:GOSUB 14100:sf=sm-m*A:et=e*n
380
390 REM TABULATED OUTPUT
400 PRINT:PRINT"MODEL PARAMETERS:   ":PRINT:PRINT
410 PRINT"Initial discharge = ";i
420 PRINT"Saturation discharge = ";qo
430 PRINT"Recession parameter (m) = ";m
440 PRINT:PRINT:GOSUB 29000:PRINT"Press RETURN to continue":
    INPUT a$:GOSUB 18000
450
460 PRINT:PRINT"WATER BALANCE: ":PRINT:PRINT"INPUT:"
470 R=4:C=22:B=4:D=2:REM format indicators
480 PRINT"Total rainfall";:A=rt:GOSUB 17200:PRINT:PRINT:
    PRINT"OUTPUT:"
490 PRINT"Total evaporation";:R=7:A=et:GOSUB 17200:PRINT
500 PRINT"Total runoff";:R=8:A=qt:GOSUB 17200:PRINT
```

(continued)

```
510 PRINT"Total throughflow";:R=9:A=tt:GOSUB 17200:PRINT
520 PRINT"Total overland flow";:R=10:A=ot:GOSUB 17200
530 PRINT:PRINT:PRINT"CHANGE IN STORAGE:":C=29
540 PRINT"Initial soil moisture deficit";:R=13:A=s:GOSUB 17200:
    PRINT
550 PRINT"Final soil moisture deficit";:R=14:A=sf:GOSUB 17200:
    PRINT
560 PRINT:PRINT:GOSUB 29000:PRINT"Press RETURN to continue":
    INPUT a$:GOSUB 18000
570
580 PRINT:PRINT"HYDROGRAPH DATA: ":PRINT
590 PRINT"DAY   RAIN  EVAP RUNOFF T'FLOW O'FLOW"
600 R=3:FOR t=1 TO n:R=R+1:C=0:B=2:A=t:GOSUB 17300
610   C=4:D=1:B=3:A=r(t):GOSUB 17200:C=10:A=e:GOSUB 17200
620   C=15:D=3:B=4:A=qd(t):GOSUB 17200:C=23:A=th(t):GOSUB 17200
630   C=33:A=of(t):GOSUB 17200:NEXT t
640 GOSUB 29000:PRINT:PRINT"Press RETURN to continue":INPUT a$:
    GOSUB 18000
650
660 REM Graphical output; use of library routines
670 GOSUB 20000:LET FG=0:REM Graphics screen - new graph
680 NX=0:XX=n:NY=0:XY=qo:rr=0:REM Sets limits of axes
690 FOR t=1 TO n:IF qd(t)>XY THEN XY=qd(t)
700   IF r(t)>rr THEN rr=r(t)
710   NEXT t:XY=INT(XY+rr+5):GOSUB 21000:GOSUB 22000
720 X=0:Y=i:GOSUB 26000:FOR t=1 TO n:X=t:Y=qd(t):GOSUB 26000:
    NEXT t:REM Draws hydrograph
730 REM Horizontal lines qo (sat) and e (evap)
740 X=n:Y=qo:GOSUB 25000:X=0:GOSUB 26000:A$="qsat":X=n-4:
    GOSUB 23000
750 IF e<>0 THEN Y=XY-e:X=0:GOSUB 25000:X=n:GOSUB 26000:
    A$="evap":X=n-4:Y=XY-e:GOSUB 23000
760 FOR t=1 TO n:IF r(t)<>0 THEN X=t-1:XA=t:Y=XY:YA=XY-r(t):
    GOSUB 27000
770   NEXT t:REM rainfall histogram
780 PRINT"    RAINFALL & RUNOFF AGAINST TIME":PRINT:
    PRINT"Press RETURN to end"
790 INPUT A$:GOSUB 29500:GOSUB 15000:GOSUB 28000:END
800
1000 DATA 5,10,15,5,2,0,0,0,0,0,0,0,0,0,0,0,0,0,0,0:
     REM 20 days of rainfall
1010
```

Figure 3.5   Program STORFLO to predict runoff

R_n is net radiation (W/m^2)
G is soil heat flux (W/m^2)
ρ is air density (kg/m^3)
c_p is specific heat of air at constant pressure (1005 J/kg)
e_s is saturation vapour pressure at screen temperature (mb)
e is screen vapour pressure (mb)
L is latent heat of vaporization (2465000 J/kg)
g is the psychromatic constant (0.66)
r_s is bulk surface resistance (s/m)
r_a is bulk aerodynamic resistance (s/m)

A program EVAP which performs these calculations and incorporates the various ways of calculating r_s is given in Figure 3.6. Typical values for each

```
 10 GOSUB 10000:REM Initialize machine
 20 REM Set up arrays - m days in months; l,s grass constants
 30 DIM m(12,2),l(12),s(12)
 40
 50 PRINT"EVAPORATION MODEL":PRINT
 60 PRINT"Based on season and crop type"
 70 PRINT"Data given is for crops in Norfolk":PRINT
 80
 90 REM Read in data for calendar and grass constants
100 RESTORE
110 FOR k=1 TO 2:FOR j=1 TO 12:READ m(j,k):NEXT j:NEXT k
120 FOR j=1 TO 12:READ l(j):NEXT j
130 FOR j=1 TO 12:READ s(j):NEXT j
140
150 REM Input date and calculate day number
160 C=0:R=5:A=1:LL=1:UL=12:C$="Month number":GOSUB 12500:mn=A
170 R=6:UL=m(mn,1):A=1:C$="Day number":GOSUB 12500:nd=A:
    d=m((mn-1),2)+nd
180
190 REM Input weather conditions
200 PRINT:PRINT:PRINT"Input weather conditions:":PRINT
210 R=11:LL=-100:UL=800:A=100:
    C$="Net radiation (-100-800 W/m**2): ":GOSUB 12510:rn=A
220 R=R+1:LL=-75:UL=55:A=10:C$="Temperature (-75-55 deg.C): ":
    GOSUB 12510:t=A
230 R=R+1:LL=0:UL=20:A=0:
    C$="Wet bulb depression (0-20 deg.C): ":GOSUB 12510:wb=A
240 R=R+1:LL=0:UL=150:A=1:
    C$="Wind speed 2 m above crop (0-150m/s): ":GOSUB 12510:u=A:
    IF u=0 THEN u=0.001
250 PRINT:PRINT"Press RETURN to continue":INPUT a$
260
270 r=1.292:cp=1010:k=0.41:REM air density (kg/m**3),
    specific heat of air (J/kg),von Karman's constant
280 ga=0.00066*(1+(0.00115*t))*982.5:REM psychromatic constant
290
300 REM Gradient of sat. vapour pressure curve
310 x=t-0.5:GOSUB 660:y1=x:x=t+0.5:GOSUB 660:y2=x:dp=y2-y1
320
330 REM Difference between actual and sat. vapour pressure
340 x=t-wb:GOSUB 660:y1=x:x=wb:GOSUB 660:y2=x-0.799*(wb):
    dt=y1-y2
350
360 REM Prints out heading for table
370 GOSUB 18000:PRINT:PRINT:PRINT"Data for central Norfolk"
380 PRINT:PRINT"Date is ";nd;"/";mn:PRINT
390 PRINT"Crop type";TAB(19);"ra";TAB(26);"rs";TAB(33);"Et"
400 PRINT TAB(19);"s/m";TAB(26);"s/m";TAB(31);"mm/hour":PRINT:
    PRINT
410
420 REM Loops through crop types given in data statements
430 READ nc:R=9:FOR j=1 TO nc:R=R+1
440    READ a$,i,h1,h2,ds,de,df,dh,lm,sc
450
460    REM Calculation of bulk aerodynamic resistance (s/m)
470    h=h2:IF de=dh THEN GOTO 500:REM Non-seasonal crop
480    IF d<de OR d>dh THEN h=0.05:GOTO 500:REM Bare soil
490    IF d<df AND df<>de THEN h=h1+(h2-h1)*(d-de)/(df-de)
500    z0=0.1*h:A=2/z0:GOSUB 14100:ra=A^2/((k^2)*u)
510
```

(continued)

```
520    REM Calculate bulk surface resistance (s/m)
530    ss=100:REM Bare soil resistance, wet conditions
540    ON i GOSUB 710,740,740,740,800
550    REM grass/cereals/roots/deciduous trees/conifers
560
570    REM Calculates evapo-transpiration in mm/hour
580    et=((dp*rn)+(r*cp*dt/ra))/(dp+ga*(1+rs/ra))/694.5
590
600    REM Prints out results
610    PRINT a$;:C=15:B=4:D=1:A=ra:GOSUB 17200
620    C=22:B=5:A=rs:GOSUB 17200:C=31:B=2:D=3:A=et:GOSUB 17200:
       PRINT
630
640    NEXT j:PRINT:PRINT:END
650
660 REM Subroutine to calculate vapour pressure
670 A=x+273.16:GOSUB 14100:
    x=EXP(54.878919-(6790.4985/(x+273.16)+5.02808*A))
680 RETURN
690
700
710 REM Resistance calculation for grass
720 la=l(mn):sc=s(mn):rs=sc*ss/(ss-(ss-sc)*0.7^la):RETURN
730
740 REM Resistance calculation for cereals + deciduous trees
750 IF d<de OR d>dh THEN rs=ss:RETURN
760 la=lm:IF d<df AND df<>de THEN
    la=(lm-0.1)*(d-de)/(df-de)+0.1
770 IF i<3 AND d>df THEN
    sc=sc+50*(d-df)/(dh-df)+500*((d-df)/(dh-df))^3
780 rs=sc*ss/(ss-(ss-sc)*0.7^la):RETURN
790
800 REM Resistance calculation for conifers
810 IF t<-5 OR dt>20 THEN sc=10000:rs=sc:RETURN
820 IF t<20 sc=25*sc/(t+5)
830 IF dt<20 sc=sc/(1-0.05*dt)
840 rs=sc:RETURN
850
860 REM Data for conversion of date to days (not a leap year)
870 DATA 31,28,31,30,31,30,31,31,30,31,30,31
880 DATA 31,59,90,120,151,181,212,243,273,304,334,365
890
900 REM Data for grass (leaf areas, resistances)
910 DATA 2,2,3,4,5,5,5,5,4,3,2.5,2
920 DATA 80,80,60,50,40,60,60,70,70,70,80,80
930
940 REM Number of crop types
950 DATA 7
960 REM Data given: Crop, Index, Min. height, Max. height
970 REM Data:Day sown,Day emerges,Day full cover,Day harvested
980 REM Data:Leaf area index,surface resistance
990 DATA Grass,1,0.15,0.15,0,0,0,0,2,40
1000 DATA Winter wheat,2,0.08,0.8,79,79,175,235,5,40
1010 DATA Spring barley,2,0.05,0.8,77,92,172,232,5,40
1020 DATA Potatoes,3,0.05,0.6,95,120,185,242,4,40
1030 DATA Sugar beet,3,0.05,0.35,85,149,186,298,4,40
1040 DATA Deciduous trees,4,0.175,10.0,108,148,290,330,6,80
1050 DATA Conifers,5,10,10,0,0,0,0,6,70
1060
```

Figure 3.6   Program EVAP to predict evapo-transpiration

56

Table 3.1 Typical weather conditions for input to evapo-transpiration model

	Likely range of values		Typical values at noon in Britain	
	Maximum	Minimum	Winter	Summer
Net radiation (W/m²)	800	−100	30	600
Temperature (°C)	55	−75	2	19
Wet bulb depression (°C)	20	0	1	4
Wind speed at 2 m (m/s)	150	0	3	3

of the variables used are given in Table 3.1. Interesting results can be obtained by looking at the seasonal variations in r_s and transpiration loss for different vegetation types. An example model run is provided in Figure 3.7.

This example again illustrates the use of a greater appreciation of the mechanics behind processes in giving a more reasonable prediction of potential evapo-transpiration and, in this case, much more detailed information. The model, however, still does not incorporate the calculation of night-time losses, nor the effect of interception loss (see Rutter *et al.*, 1971), nor the feedback with soil moisture. To perform this last function, the program presented here needs to incorporate a runoff model similar to that developed in the previous section.

3.4 ANALYSIS OF EROSION PROCESSES AND A SIMPLE LANDSLIDE MODEL

So far, two analogies have provided a conceptual basis for modelling the movement of water. When considering the movement of sediment, it is easy enough to provide a physical basis to the calculations from a knowledge of simple physics. In this section, the movement of a single grain in a fluid flow is examined. Movement is assumed to occur as a slide and the same type of analysis is adapted to predict the occurrence of (much simplified) dry landslides.

Essentially, most problems concerning the initiation of motion can be thought of in terms of a balance of the forces operating (Carson and Kirkby, 1972). For the example of river gravels, the forces involved can be specified as shown in Figure 3.8 and, in the case of spherical grains in a turbulent flow, expressed as follows:

$$F_D = 0.05\rho\pi D^2 u_y^2 \tag{3.8}$$

where F_D is the downstream drag force (N/m²)
         ρ is the density of water (kg/m³)
         D is the grain diameter (m)
         u_y is the instantaneous downstream velocity at the grain centre (m/s)

```
                    EVAPORATION MODEL

                    Based on season and crop type
                    Data given is for crops in Norfolk

                    Enter Month number (1-12): 6
                    Enter Day number (1-30): 12


                    Input weather conditions:

                    Net radiation (-100-800 W/m**2): 500
                    Temperature (-75-55 deg.C): 16
                    Wet bulb depression (0-20 deg.C): 3
                    Wind speed 2 m above crop (0-150m/s): 1

                    Press RETURN to continue
                    ?



                    Data for central Norfolk

                    Date is 12/6

                    Crop type            ra      rs     Et
                                         s/m     s/m   mm/hour


                    Grass              142.4    64.3   0.455
                    Winter wheat        66.3    45.7   0.489
                    Spring barley       66.0    45.6   0.490
                    Potatoes            89.5    52.0   0.472
                    Sugar beet         137.4    60.8   0.459
                    Deciduous trees     46.7    94.6   0.389
                    Conifers             2.9   163.4   0.184
```

Figure 3.7   Example run of program EVAP to predict evapo-transpiration

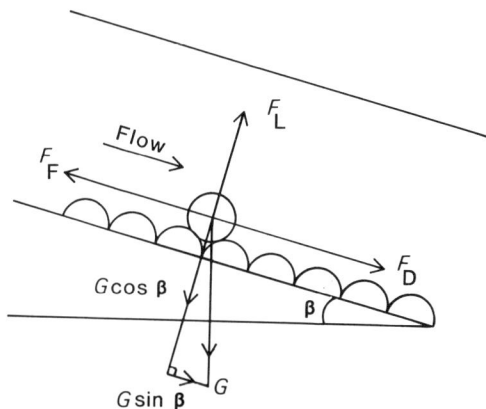

Figure 3.8   Forces acting on a single spherical grain in a fluid flow

$$F_L = 0.05\, C_L\, \rho\pi\, D^2\, u_y^2 \tag{3.9}$$

where C_L is a coefficient of lift (Naden, 1987)

$$G = (\rho_s - \rho)\, g\left(\frac{\pi}{6}\right)D^3 \tag{3.10}$$

where ρ_s is the density of sediment (kg/m^3)
g is acceleration due to gravity (m/s^2)

$$F_F = (G\cos\beta - F_L)\tan\phi \tag{3.11}$$

where β is the average bedslope (degrees)
ϕ is the angle of friction (degrees)

The balance of forces is thus given by

$$F_D = (G\cos\beta - F_L)\tan\phi - G\sin\beta \tag{3.12}$$

A very simple program, EROSION, to calculate the instantaneous down-stream velocity required to move a grain is given in Figure 3.9 with a model run shown in Figure 3.10. Results are given for a series of grains protruding from the bed by different amounts (see Fenton and Abbott, 1977) – a grain resting on top of a bed of equal-sized spheres has a relative protrusion of 0.88; a grain within the bed of 0. The program also calculates the mean velocity at the grain centre based on the logarithmic flow law:

$$u_y = 5.75\, u_* \log\left(30.1\,\frac{y}{D}\right) \tag{3.13}$$

where u_y is the mean velocity at height y (m/s)
u_* is bed shear velocity (m/s)
y is height above the bed (m)

and its standard deviation based on the equation (Naden, 1987):

$$\sigma_u = 0.16\, u_y \left(\frac{y}{D}\right)^{-0.65} \tag{3.14}$$

Users of the program are left to decide for themselves which of the grains are likely to move, although the program could be formally developed into a stochastic model (Chapter 5) giving the probabilities of grain movement.

The balance of forces idea has also been applied to the case of very simple dry wedge landslides. The diagram of the landslide and the forces involved are described in Figure 3.11. In this case, the weight of the projected landslide (represented by the triangle ABC in Figure 3.11) has to be balanced against the forces of friction and cohesion acting along the length of the slip slope (BC in Figure 3.11), i.e.

$$W\sin a = W\cos a \tan\phi + cL \tag{3.15}$$

```
 10 GOSUB 10000:REM Initialize machine
 20 PRINT:PRINT"Program calculates threshold of motion":
    PRINT"for an individual grain in a fluid flow":PRINT
 30
 40 PRINT"Input parameters:":PRINT
 50 R=6:C=0:LL=0.001:UL=1:A=0.1:C$="Grain size (0.001-1m) = ":
    GOSUB 12510:g=A
 60 R=7:LL=0:UL=0.1:A=0:C$="Average bed slope (0-0.1) = ":
    GOSUB 12510:b=ATN(A)
 70 R=8:LL=0.1:UL=5:A=2:C$="Mean flow velocity (0.1-5m/s) = ":
    GOSUB 12510:v=A
 80 R=9:LL=0.1:UL=10:A=1:C$="Depth of flow (0.1-10m) = ":
    GOSUB 12510:d=A
 90
100 dg=1.65*9.81*3.141592653/6*g:c=COS(b):s=SIN(b):
    REM constants
110 A=30.1*0.4*d/g:GOSUB 14200:us=v/A:REM bed shear velocity
120
130 PRINT:PRINT" Relative           Velocities":
    PRINT"protrusion   threshold   mean   st.dev":
    PRINT:REM title
140 R=13:B=2:D=2:REM sets up format for table
150
160 REM Calculation of velocities (three basic geometries)
170 FOR y=0.88 TO 0.0 STEP -0.1:yy=y:IF y<0.35 THEN yy=0.35
180  IF y<>0 THEN p=SQR(1-y*y)/y:REM tan(friction angle)
190  A=30.1*yy:GOSUB 14200:uy=us*A:REM velocity at grain
     centre
200  su=0.16*uy*(yy^-0.65):REM standard deviation
210  IF y>0.35 THEN u=uy+(-uy+SQR(uy^2-(1+0.593*p)*
     (uy^2-dg/0.157*(c*p-s))))/(1+0.593*p)
220  IF y<=0.35 THEN u=SQR(dg*c/0.07):IF y>0 THEN
     ul=SQR(dg*(c*p-s)/(0.12+0.07*p)):IF ul>u THEN u=ul
230  R=R+1:C=2:A=y:GOSUB 17200:C=15:A=u:GOSUB 17200:C=24:
     A=uy:GOSUB 17200:C=32:A=su:GOSUB 17200
240
250  NEXT y:PRINT:PRINT
260 END
270
```

Figure 3.9 Program EROSION to calculate the velocity of flow required to move a single spherical grain

where W is weight of landslide ABC per unit slope width (N/m)
       a is angle of slip (degrees)
       ϕ is angle of friction (degrees)
       c is cohesion of material (N/m^2)
       L is length of slip slope BC (m)

The weight of the landslide is calculated by multiplying the unit weight of material by the volume of the slide as given by simple geometry:

$$W = 0.5\gamma H^2 \frac{\sin(i-a)}{\sin a \sin i} \qquad (3.16)$$

where γ is unit weight of material (N/m^3)
      H is height of cliff (m)

```
Program calculates threshold of motion
for an individual grain in a fluid flow

Input parameters:

Grain size (0.001-1m) = 0.1
Average bed slope (0-0.1) = 0.001
Mean flow velocity (0.1-5m/s) = 2
Depth of flow (0.1-10m) = 1

Relative              Velocities
protrusion    threshold    mean    st.dev

   0.88          1.70       1.37     0.24
   0.78          2.02       1.32     0.25
   0.68          2.27       1.26     0.26
   0.58          2.49       1.19     0.27
   0.48          2.68       1.11     0.29
   0.38          2.86       1.02     0.31
   0.28          3.48       0.98     0.31
   0.18          3.48       0.98     0.31
   0.08          3.48       0.98     0.31
```

Figure 3.10 Output from program EROSION to predict the velocity of flow required
to move a single grain

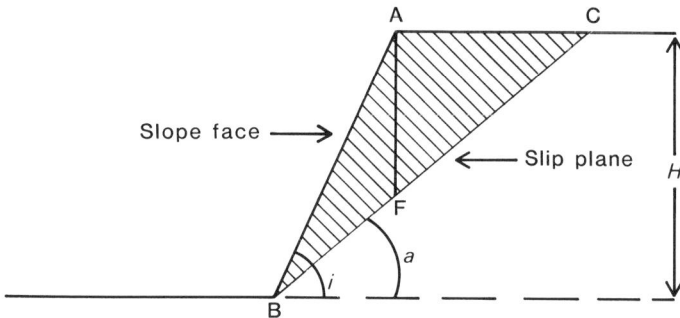

Figure 3.11 Definition diagram for a simple wedge landslide; i is slope angle, a is
slip angle.

Table 3.2 gives suitable values of the angle of friction and cohesion per unit
weight of material for use in the model. Although the dry landslide analysed
here is a very simple situation (for more complicated models see Milligan
and Houlsby, 1984), it can be useful to introduce the concept of a balance
of forces on a large scale and the idea of hillside stability as represented by
the safety factor.

$$SF = \frac{\text{forces resisting movement}}{\text{forces promoting movement}} \qquad (3.17)$$

Table 3.2 Characteristics of geological materials for use in landslide model (based on Statham, 1977)

Material	Angle of internal friction (degrees)	Cohesion ($\times 10^3$ N/m^2)	Unit weight of material ($\times 10^3$ N/m^2)	Relative cohesion (c/γ)
Boulder clay	32	8	19–22.5	0.4
Marine clay	27	10	19–22.5	0.5
Coal measures	25	7	19–22.5	0.3
Lias clay	23	7	19–22.5	0.8
Clay shales	20	9	19–22.5	0.4
London clay	25	12	19–22.5	0.6
Granite	30–50	100–300	26.1	3.8–11.5
Sandstone	30–45	50–150	19.5	2.6– 7.7
Shale	27–45	20–100	24.0	0.8– 4.2
Limestone	30–50	25–100	31.7	0.8– 3.2
Quartzite	30–50	100–300	26.1	3.8–11.5

For values of the safety factor greater than 1, the forces resisting movement are greater than the forces promoting movement and a landslide will not occur. For safety factors less than 1, it is useful to know which is the most likely landslide to occur. The value of the slip angle in these equations, however, is unknown and the program, SLIDE (Figure 3.12), therefore, calculates the safety factor for a range of angles from 0 to the angle of the slope face. For the smallest of these safety factors, the program then calculates the scale (in square metres) of the most likely landslide. Thus, the program can also be used to optimize the design of a cutting.

3.5 ECOLOGICAL MODELS

A final example demonstrating the range of process models used in physical geography is given in the biological field, where organisms can multiply, grow and die; eat and be eaten.

The equations governing such processes are very simple. Considering, first, a single population, then the intrinsic growth rate of the population is given by the exponential growth curve

$$\frac{dN}{dt} = rN \tag{3.18}$$

where N is the number of individuals in the population
r is the intrinsic rate of growth
t is time

```
 10 GOSUB 10000:REM Initialize machine
 20 R=1: C=7:A$="WEDGE LANDSLIDE ANALYSIS":GOSUB 16000:
    REM Title
 30 RESTORE:READ ph,hc,h,g:REM Values offered initially
 40 DIM sf(90):REM Array of safety factors
 50 r=3.14159/180:REM conversion degrees to radians
 60
 70 REM Input material properties
 80 C=0:R=3:A$="Angle of friction in degrees:":GOSUB 16000:C=2:
    R=4:A$="Clays and shales at 10 to 25 deg.":GOSUB 16000
 90 R=5:A$="Sands and gravels at 30 to 40 deg":GOSUB 16000:R=6:
    A$="Unweathered hard rock at 50 to 80 deg":GOSUB 16000
100 R=7:LL=10:UL=60:A=ph:C$="Angle":GOSUB 12500:ph=A:
    QH=TAN(ph*r)
110 C=0:R=9:A$="Cohesion as max. cliff height (metres)":
    GOSUB 16000:R=10:C=2:
    A$="Clays @ 5 to 15m: Sands & gravels @ 0":GOSUB 16000:
    R=11:A$="Unweathered hard rock at 50 to 1000m":GOSUB 16000
120 LL=0:UL=1000:A=hc:R=12:C$="Critical height":GOSUB 12500:
    hc=A:co=A/4/TAN(r*(45+g/2))
130
140 REM Input slope dimensions
150 R=14:C=0:C$="Vertical height(m)":LL=1:UL=1000:A=h:
    GOSUB 12500:h=A
160 R=15:LL=10:UL=90:A=g:C$="Angle of face (deg)":GOSUB 12500:
    g=A:gr=g*r
170 GOSUB 20000:NX=0:NY=0:XY=h:XX=1280/1024*XY:
    IF g=90 THEN GOTO 190
180 QG=TAN(gr):U=XY/QG/XX/.6:IF U>1 THEN XX=XX*U:XY=XY*U
190 GOSUB 21000:FG=0:X=NX:Y=0:GOSUB 26000:X=0:GOSUB 26000
200 IF g<>90 THEN X=h/QG
210 Y=h:GOSUB 26000:X=XX:GOSUB 26000
220 A$="Looking for lowest Safety Factor":GOSUB 24000
230
240 REM Calculates safety factors for different slip angles
250 im=g-1:s=100:ic=9:FOR i=1 TO im:ir=i*r
260   w=h*h/2*SIN((g-i)*r)/SIN(gr)
270   QI=TAN(ir):sf(i)=QH/QI+co/SIN(ir)*h/w
280   IF sf(i)>=s THEN GOTO 310
290   s=sf(i):R=1:C=0:A$="SF ="+STR$(s)+"      ":GOSUB 16000:
      ic=i
300   FG=0:X=0:Y=0:GOSUB 26000:Y=h:X=Y/QI:GOSUB 26000
310   IF i-ic>5 THEN im=i:i=g
320   NEXT i
330
340
350 R=0:C=0:A$="":GOSUB 24000:GOSUB 19010:GOSUB 15000:
    REM clear to text screen
360 REM Prints out table of safety factors
370 A$="SLOPE HT = "+STR$(h):GOSUB 16000:R=1:
    A$="CRIT HT  = "+STR$(hc):GOSUB 16000
380 C=20:R=0:A$="FACE ANGLE(deg)= "+STR$(g):GOSUB 16000:
    R=1:A$="FRICTION ANGLE = "+STR$(ph):GOSUB 16000
390 A$="=====":FOR I=1 TO 3:A$=A$+A$:NEXT I:R=2:C=0:GOSUB 16000
400 R=3:C=0:A$="  Slip angle        Safety Factor":GOSUB 16000
410 mn=im-10:IF mn<1 THEN mn=1
420 R=4:FOR i=mn TO im:R=R+1
430   C=3:B=5:A=i:GOSUB 17300:C=22:B=2:D=3:A=sf(i):GOSUB 17200
440   IF i=ic THEN C=18:A$=">>>":GOSUB 16000
450   NEXT i:R=17
460
470 C=0:IF s>1 THEN A$="A landslide will not occur":
    GOSUB 16000:GOTO 530
```

(continued)

```
480 A$="Cross-sectional area of the most likely slide:":
    GOSUB 16000
490 PRINT:w=h*h/2*SIN((g-ic)*r)/SIN(gr):v=w/SIN(ic*r)
500 R=18:B=4:C=16:A=v:GOSUB 17300:C=24:A$="sq.m":GOSUB 16000
510 C=0:R=20:A$="on a slide plane at":GOSUB 16000:B=2:C=21:
    A=ic:GOSUB17300:C=25:A$="degrees":GOSUB 16000
520 C=0:R=21:A$="with a safety factor of":GOSUB 16000:C=25:B=2:
    D=3:A=s:GOSUB 17200
530 GOSUB 29000:C=0:R=23:C$="Another go?: Y/N >":A$="Y":
    GOSUB 11000:IF A$="N" THEN END
540 GOSUB 18000:GOTO 70
550
560 REM Data giving initial values for angle of friction,
    critical ht, actual ht and face angle
570 DATA 10,50,100,55
```

Figure 3.12   Program SLIDE to predict a simple dry wedge landslide

However, populations are limited by the amount of resources and space available to them and so it is necessary to define the carrying capacity of the ecosystem for a particular species, i.e. the maximum number of individuals which the ecosystem can support. Incorporating this idea into equation (3.18) gives the typical logistic curve for population growth (Kormondy, 1984):

$$\frac{dN}{dt} = rN\frac{K-N}{K} \tag{3.19}$$

where K is the carrying capacity. However, species do not exist in isolation. There is competition between species for resources and the important relationship between predators and prey which is the one that will be explored here. One of the ways in which the predator–prey relationship can be modelled is using the Lotka–Volterra equations (Hassell, 1976). For a single predator species and a single prey species, the pair of simultaneous equations can be expressed as

$$\frac{dN}{dt} = rN - c_1NP \tag{3.20}$$

and

$$\frac{dP}{dt} = c_2NP - dP \tag{3.21}$$

where N is the number of prey
         r is the intrinsic growth rate of the prey
         c_1 is the coefficient of attack
         P is the number of predators
         d is the death rate of predators in the absence of prey
         c_2 is the prey–predator conversion rate

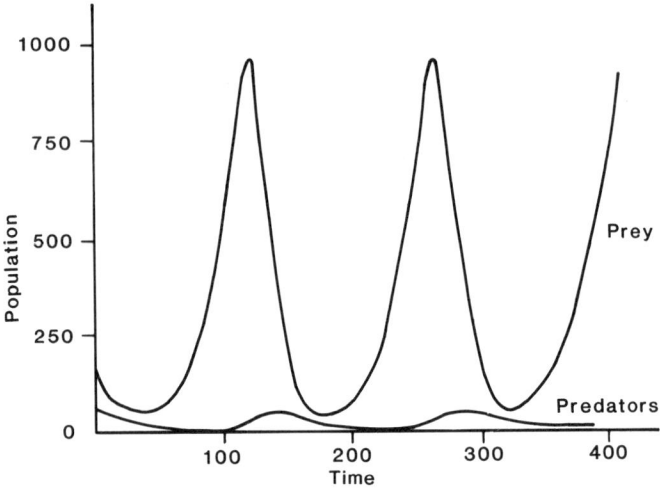

Figure 3.13   Illustration of the Lotka–Volterra equations ($n=200$, $p=50$, $r=8$, $d=3$, $c=0$)

These equations, then, simply represent the intrinsic rate of growth of the prey minus the number of prey eaten by predators and the growth rate of the predator population as controlled by the food supply of prey minus the number of predators which starve. The solution of these equations is the characteristic out-of-phase oscillatory pattern of prey and predator populations (Figure 3.13). This is in many cases an unrealistic picture of population dynamics and in the model given in the computer program ECOLOGY (Figure 3.14) a more refined version which includes the idea of carrying capacity (variables a and b in the first data statement are simply the reciprocals of K) and the availability of hiding places for the prey. A typical model run ending in equilibrium prey and predator populations is given in Figure 3.15.

The overall performance of the model is extremely sensitive to small variations in the input parameters. However, with care, the program can be used to illustrate all the functions discussed in this section. Equation (3.18) can be explored by having no predators and varying the value of r while equation (3.19) can be investigated by changing the value of a in the data statement 1000. The results of the original Lotka–Volterra equations are reproduced if a and b remain zero and no cover is provided for the prey. This program, therefore, provides a very versatile means of investigating population dynamics.

3.6   CONCLUSION

This chapter has introduced a range of process-based models. These have been related to a number of very basic concepts which lie behind many of

```
  10 REM Simple predator-prey model based on the Lotka-Volterra
     equations
  20 REM assumes population age and sex structure stable;
     migration negligible
  30
  40 GOSUB 10000:REM initialize machine
  50
  60 RESTORE:READ c1,c2,a,b:REM Parameters given in data
     statements
  70 REM c1 is coefficient of attack; c2 is prey-predator
     conversion rate
  80 REM a and b are coefficients of intraspecific competition
  90 READ t1,t2,dt:REM Time data - start, finish, increment
 100
 110 PRINT:PRINT"Predator-prey populations over time"
 120 PRINT:PRINT:PRINT"Input initial populations:"
 130 PRINT:PRINT"Enter a number between 0 and 400   ":C=0:R=7:
     LL=0:UL=400:A=200:C$="Initial population of prey = ":
     GOSUB 12510:n=A
 140 R=8:A=50:C$="Initial population of predators = ":
     GOSUB 12510:p=A
 150 PRINT:PRINT:PRINT"Input controlling parameters:":PRINT
 160 PRINT"Enter a number between 1 and 10   ":R=14:UL=10:A=8:
     C$="Intrinsic rate of increase of prey = ":GOSUB 12510:
     r=A/100
 170 R=15:A=3:C$="Death rate of predators if no prey = ":
     GOSUB 12510:d=A/100
 180 PRINT:PRINT"Enter a number between 0 and 400   ":R=18:
     UL=400:A=0:      C$="Cover afforded to prey = ":
     GOSUB 12510:c=A
 190 PRINT:PRINT:PRINT"Press RETURN to continue   ":INPUT a$
 200
 210 GOSUB 20000:REM Sets up graphics screen with axes
 220 FG=0:NX=t1:XX=t2:NY=0:XY=1000:GOSUB 21000:GOSUB 22000
 230
 240 PRINT TAB(6);"POPULATION VERSUS TIME"
 250 PRINT"Initial popn:   ";n;" prey   ";p;" predators"
 260 IF c<>0 THEN X=t1:Y=c:GOSUB 26000:X=t2:GOSUB 26000:
     A$="COVER":X=0.75*t2:Y=Y+50:GOSUB 23000:REM plots cover if
     present
 270 REM Calculates populations over time and plots time-graph
 280 FOR t=t1 TO t2 STEP dt
 290   n=n+((r-a*r*n)*n)*dt:IF n>c THEN n=n-(c1*p*(n-c))*dt
 300   p=p-((b*p+d)*p)*dt:IF n>c THEN p=p+(p*c2*(n-c))*dt
 310   IF n=0 AND p=0 THEN t=t2
 320   X=t:Y=n:GOSUB 25000:Y=p:GOSUB 25000
 330   NEXT t
 340
 350 PRINT"Final popn:     ";INT(n);" prey  ";INT(p);" predators"
 360 PRINT"Press RETURN to end":INPUT A$:GOSUB 29500:
     GOSUB 15000:GOSUB 28000:END
 370
1000 DATA 0.003,0.0001,0.0,0.0:REM Model parameters given
1010 DATA 0,400,1:REM Time data
```

Figure 3.14   Program ECOLOGY to explore population dynamics

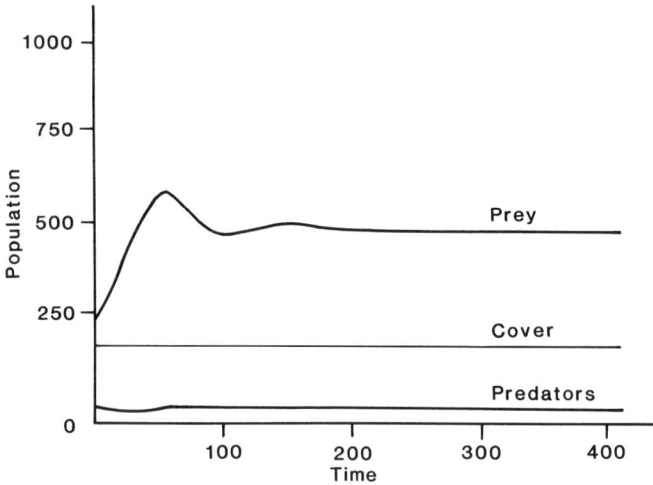

Figure 3.15 Typical run of ECOLOGY model showing equilibrium populations ($n=200$, $p=50$, $r=8$, $d=3$, $c=200$)

the processes in physical geography: storage-flow equations, a resistance analogy, balance of forces and growth/decay functions. Coupled with the black box models given above, these process ideas form powerful tools or input to the mass balance models developed in Chapter 4.

3.7 REFERENCES

Carson, M. A. and Kirkby, M. J. (1972) *Hillslope Form and Process*. Cambridge University Press, Cambridge.

Clark, R. T. and Newson, M. D. (1978) Some detailed water balance studies of research catchments, *Proceedings of the Royal Society of London, series A*, **363**, 21–42.

Dunne, T. and Black, R. D. (1970) An experimental investigation of runoff production in permeable soils, *Water Resources Research*, **6**, 478–90.

Dunne, T. and Leopold, L. B. (1978) *Water in Environmental Planning*, W. H. Freeman and Co., San Francisco.

Fenton, J. D. and Abbott, J. E. (1977) Initial movement of grains on a stream bed: the effect of relative protrusion, *Proceedings of the Royal Society of London, series A*, **352**, 523–37.

Hassell, M. P. (1976) *The Dynamics of Competition and Predation*. Edward Arnold, London.

Kirkby, M. J. (1975) Hydrograph modelling strategies. In R. F. Peel, M. D. Chisholm and P. Haggett (eds.) *Processes in Physical and Human Geography*, Academic Press, London, 69–90.

Kormondy, E. J. (1984) *Concepts of Ecology* (3rd edn), Prentice-Hall, New Jersey.

Meteorological Office (1981) *The Meteorological Office Rainfall and Evaporation Calculation System MORECS*, Hydrological Memorandum No. 45.

Milligan, G. W. E. and Houlsby, G. T. (1984) *BASIC Soil Mechanics*, Butterworths, London.

Monteith, J. L. (1973) *Principles of Environmental Physics*, Edward Arnold, London.

Naden, P. S. (1987) An erosion criterion for gravel-bed rivers. *Earth Surface Processes and Landforms*, **12**, 83–94.

Natural Environment Research Council (1975) *Flood Studies Report*, Natural Environment Research Council, UK.

Penman, H. L. (1948) Natural evaporation from open water, bare soil and grass, *Proceedings of the Royal Society, series A*, **193**, 120–45.

Rutter, A. J., Kershaw, K. A., Robins, P. C. and Morton, A. J. (1971) A predictive model of rainfall interception in forests, *Agricultural Meteorology*, **9**, 856–61.

Statham, I. (1977) *Earth Surface Sediment Transport*, Clarendon Press, Oxford.

Ward, R. C. (1975) *Principles of Hydrology* (2nd edn), McGraw-Hill, London.

Weyman, D. R. (1974) Runoff process, contributing area and streamflow in a small upland catchment. In K. J. Gregory and D. E. Walling (eds) *Fluvial Processes in Instrumented Watersheds*, Special Publication No. 6, Institute of British Geographers, 33–43.

CHAPTER 4

Mass and Energy Balance Models

4.1 INTRODUCTION

In many problems of interest to physical geographers, a budget of water, sediment or energy provides a useful framework within which process models can be fitted. A budget is not only of central interest in many cases, but it provides a well-founded physical basis which holds a model together. In many of the hydrological examples discussed above, a water balance is a fundamental constraint on detailed processes. In the catchment model (STORFLO) in Chapter 3, the process computations are based on the assumption that rainfall must contribute to runoff, evapo-transpiration or soil moisture: this is a simple water budget, and examples at this and more complex levels will be explored explicitly in this chapter.

All mass or energy budgets can be reduced to a simple storage equation, and it is helpful to begin from this form:

$$\text{Inflow} - \text{Outflow} = \text{Net increase in storage} \qquad (4.1)$$

This storage equation relies on the physical principles of conservation of mass and energy. These principles should be used thoughtfully, to make sure that all flows and storages are accounted for; or, if neglected, then neglected knowing that the contribution is too small to matter for the particular problem of interest. There are absolute constraints on mass or energy conservation, because nuclear reactions convert mass into energy or vice versa, but these changes can usually be neglected as very very small.

Barring nuclear reactions, it is justifiable to budget not only total mass, but mass of individual chemical elements (e.g. hydrogen, carbon, silicon), and usually of compound substances like water, or total sediment provided that allowance is made for changes like those involved in evapo-transpiration or solution. Mass budgets are therefore fundamental to most hydrological models with any physical basis, and to physically based models for hillslope or soil evolution. They may also be used in the context of nutrient cycling (budgeting carbon or major nutrients) or ecology (budgeting biological populations).

Although energy budgets are, in principle, equally universal, in practice many processes use energy very inefficiently, with large losses as heat and

68

friction. As an example, sediment transport in rivers is thought to use at most 5 per cent of a river's total energy, so that the use of an energy budget to calculate sediment load involves estimating both the total energy of the river, and the total frictional loss. Inaccuracies of 5 per cent in either of these components can lead to greatly exaggerated errors in the energy available for sediment transport, which therefore cannot be reliably estimated from an energy budget, even though the principle of energy conservation remains valid. The one area where an energy budget is important is in microclimatology. Here the sun's radiation is divided between a number of uses, and there is little or no loss involved. This balance is perhaps most important to physical geographers for estimating evapo-transpiration and near ground temperatures, and is implicit in the model EVAP described in Section 3.2.

In setting up a mass or energy balance model, a flow or systems diagram should be drawn. A common currency must be used for the diagram, which might be energy, water, carbon, etc. The arrows represent flows of this currency, and the boxes stores, usually though not always at fixed locations. Figure 4.1 shows schematic flow diagrams for the various kinds of mass or energy balance models, as developed in this chapter. The simplest type of model is one where the currency is simply divided at each branch point (or *node*), with no storage at the nodes (Figure 4.1(a)). Flows of electricity around circuits, traffic at intersections (ideally at any rate) or rivers at tributary junctions are of this type. The simplest models with storage have a single store, or two stores (Figure 4.1(b)). For the single store, its contents are determined by the single (or multiple) inflows and outflows. For two stores, the outflow from the first store becomes the inflow to the second, and the storage equation (4.1) may be applied to each store in turn.

The logical extension is to a chain of stores in a line, the outflow from each providing the inflow to the next (Figure 4.1(c)). Usually, though not necessarily, the stores are identical in properties, and represent successive sections down a slope profile or river channel, or successive layers within the soil. In these cases the two end stores are different from the others, and have to be treated as slightly special cases because they either lack inflow or outflow; or else their inflow/outflow is controlled by external factors. The next logical extension is to a chain of stores extending in two dimensions, usually representing N–S and E–W grid cells, with the possibility of flows in both directions (Figure 4.1(d)).

Mass or energy balance models, usually combined with process models help to overcome some of the problems associated with the black box approach outlined in Chapter 2:

1 Black box models that have been empirically derived, particularly through linear regression, are often not readily transferable. For example, a model developed to predict storm discharge in one catchment often fails to predict discharge in another catchment unless the whole regression process is carried out again with data from the second catchment.

70

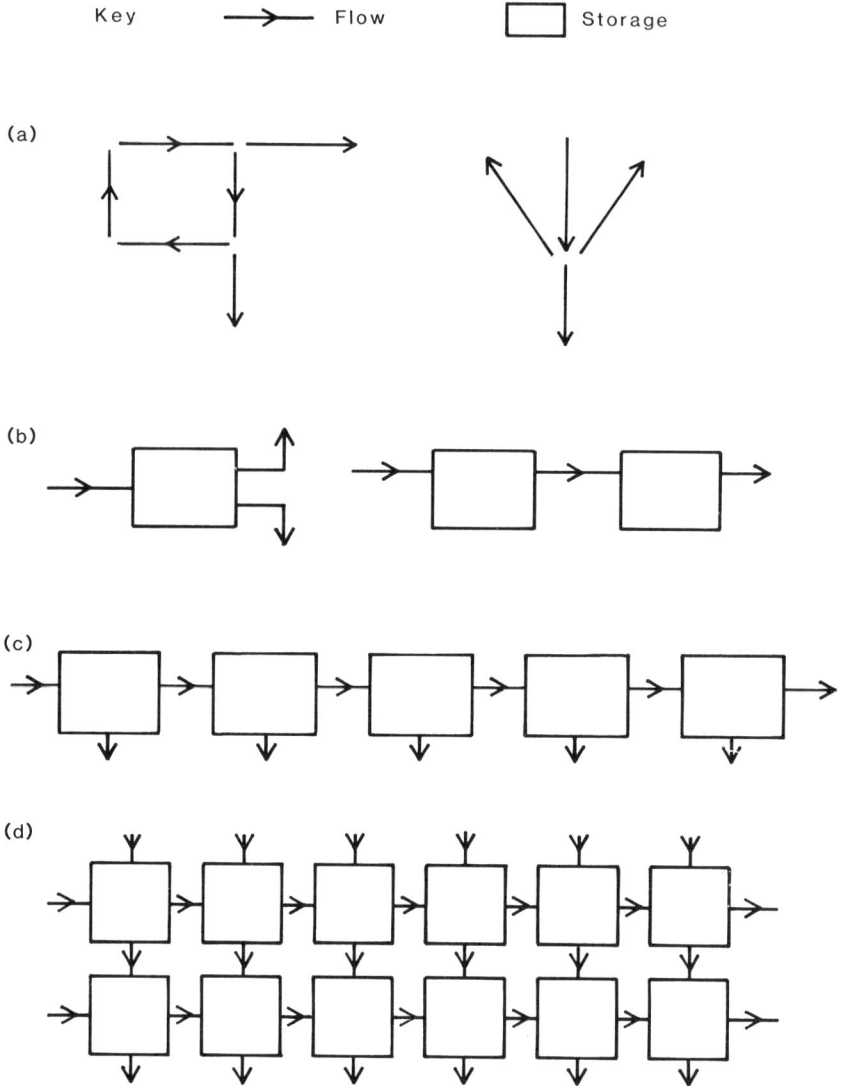

Figure 4.1 Types of mass balance models (a) models with no storage (Section 4.2); (b) one- and two-store models (Section 4.3); (c) linear multi-store models (Section 4.4); (d) two-dimensional multi-store models (Section 4.5)

2 Regression models such as those described in Section 2.2 rely on a great deal of data being available, spanning the range of conditions for which forecasts are required. For example the prediction of a flood with a particular return period requires a long record of previous flood events.

3 Black box models, by definition, give no knowledge or understanding of the actual processes occurring within them. This is not necessarily a

fault: the unit hydrograph technique (Section 2.3.2) gives no understanding of internal catchment processes, and yet has long been a successful forecasting model, particularly within large catchments. It is a well-known paradox that we may be able to predict an event or outcome very accurately without any understanding of internal process. Sunspot cycles for example may be predicted very successfully without a detailed understanding of solar physics. Successful black box methods should not be spurned, but there is long-term value in developing models that are capable of simulating the actual processes that occur as well as their final outcome; and in trying to understand why a successful black box model works so well!

4   The prediction of actual processes has the advantage that the model is potentially capable of dealing with any long-term shifts in parameter values, through for example land use changes or climatic change.

5   Finally black box models derived through regression are at best curvilinear in the relationships they use, whereas an analysis of process may suggest thresholds, reversals and other complex relationships.

Although most of these difficulties inherent in black box models are related to the processes involved, the construction of a complete model depends equally on its mass or energy balance skeleton and on the process flesh which clothes and moves it. The remainder of the chapter is set out in terms of the sequence illustrated in Figure 4.1, and the equations are all variants of the basic storage equation (4.1).

4.2 MODELS WITH NO STORAGE

If the nodes of a network, or the boxes in a systems diagram, have no storage, then the right-hand side of equation (4.1) is zero, and the equation reduces to a statement that the sum of inflows must balance the sum of outflows at every node. This is the same as Kirchoff's first law for an electrical circuit network. The nodes need not be physically identified like the connection points in a circuit, but may simply represent locations at which mass or energy is partitioned: for example the division of solar radiation at the ground surface. This type of model is applied here to two simple cases: first to the mixing of solutes and water at a tributary junction, and second to the partition of solar radiation from a cloudless sky.

4.2.1 A chemical mixing model

At a tributary junction, or where a hillside seep flows into a channel, then a mass budget can be written down for both water and total solutes. For the water

$$Q_t = Q_1 + Q_2 \qquad (4.2)$$

and for the solutes

$$C_t Q_t = C_1 Q_1 + C_2 Q_2 \qquad (4.3)$$

where Q is the water flow,
           C is the concentration of solutes
           1,2,t refer respectively to the two tributary inflows and the combined total outflow.

Pilgrim *et al.* (1978) use the model to distinguish the mix of overland flow and groundwater, or stormflow and baseflow. Its value lies in providing inferences about runoff processes in catchments where direct process observations have not been carried out. Anderson and Burt (1978) use the model to provide predictions of throughflow and Hortonian overland flow. The program listing (MIXING) is given in Figure 4.2, and an example run in Figure 4.3. The model, although simple, is capable of giving a generalized picture which demonstrates the basic budgeting technique that characterizes this type of model. For a more sophisticated model, it would be necessary to consider the processes generating the solutes, and possibly the effect of stochastic input sequences (Chapter 5).

4.2.2 A microclimatological energy balance model

At the ground surface, the various components of atmospheric energy must exactly balance because there is no storage available. Incoming (short wave) solar radiation is partly reflected according to the reflecting power (albedo) of the surface, and the remainder is used to provide long-wave back radiation, heating of the air and the ground, and energy for evapo-transpiration. The program, RADBAL, which is listed in Figure 4.4, calculates the components of this balance for clear sky conditions in order to estimate the variations in surface temperature and evapo-transpiration for a given sequence of 15-hourly solar radiation values, which are given in a DATA statement at line 460.

Back radiation is calculated from the Stefan–Boltzmann law in terms of (absolute temperature)^4. Heating of the air and the ground are treated as directly proportional to the difference between surface temperature and fixed air and ground temperatures respectively. The rate of air heating by convection is taken to be about ten times greater than the rate of ground heating by conduction. The air and ground temperatures can be chosen in the model run, and are held fixed to simplify the model. Finally the ratio of energy used in air heating to that used in evapo-transpiration from a wet surface (Bowen's ratio) is treated as a function of the difference between air and surface temperatures, on the basis of simplified theory and some empirical data. The relevant microclimatology is more fully set out in, for example, Lockwood (1979). All of these uses of energy may be expressed in terms of surface temperature, so that the requirement of total energy

```
 10 GOSUB 10000:REM initalize values
 20 C$="Enter name of Source A: ":A$="Throughflow":C=0:R=1:
    GOSUB 11000:P$=A$
 30 C$="Enter conc of "+P$+": ":A=800:R=2:GOSUB 12000: CP=A
 40 C$="Enter name of Source B: ":A$="Overland flow":C=0:R=4:
    GOSUB 11000:Q$=A$
 50 C$="Enter conc of "+Q$+": ":A=100:R=5: GOSUB 12000: CQ=A
 60 NC=CP: XC=CQ: IF CP>CQ THEN NC=CQ: XC=CP
 70 REM =====================
 80 REM Main data input and analysis
 90 N=100:DIM QQ(N),CO(N),QA(N),QB(N): YMAX=0
100 R=7: A$="Enter combined disch(Q) and Conc(C)":GOSUB 16000
110 R=8: C=5:A$="or minus values to end input":GOSUB 16000
120 I=0: R=10 : Q=100: CC=(CP+CQ)/2
130 C=0:C$=STR$(I)+": Q=":A=Q:GOSUB 12000:Q=A
140 IF A<0 THEN GOTO 250
150 C=10: C$=": and C=":A=CC:GOSUB 12000: CC=A
160 IF CC<0 THEN GOTO 250
170 IF CC<NC OR CC>XC THEN GOTO 150
180 QQ(I)=Q: CO(I)=CC
190 QB(I)=Q*(CC-CP)/(CQ-CP): QA(I)=Q-QB(I)
200 IF QQ(I)>YMAX THEN YMAX=QQ(I)
210 C=26: A$="QA="+STR$(INT(QA(I)+0.5)):GOSUB 16000
220 C=33: A$="QB="+STR$(INT(QB(I)+0.5)):GOSUB 16000
230 IF R=21 THEN GOSUB 29000
240 GOSUB 19000:I=I+1: IF I<=N THEN GOTO 130
250 GOSUB 29000:GOSUB 19010
260 REM ====================
270 REM: Draw Graphs
280 NJ=I-1:GOSUB 20000: FG=0
290 NX=0:XX=NJ*1.3 : IF XX<2 THEN XX=2
300 NY=-YMAX/2: XY=YMAX: GOSUB 21000: GOSUB 22000
310 X=0:Y=XY*0.9:A$="Disch": GOSUB 23000
320 X=XX/3: Y=XY/8:A$="Sample No":GOSUB 23000
330 A$="Rel":X=NJ/50:Y=0.4*NY:GOSUB 23000
340 A$="Conc":Y=0.55*NY:GOSUB 23000
350 FOR I=0 TO NJ
360   X=I:Y=QQ(I) :GOSUB 26000:NEXT I
370 IF NJ=0 THEN X=X+1: GOSUB 26000
380 A$="TOT": GOSUB 23000
390 FG=0: FOR I=0 TO NJ
400   X=I: Y=QA(I):GOSUB 26000: NEXT I
410 IF NJ=0 THEN X=X+1: GOSUB 26000
420 A$=LEFT$(P$,4): GOSUB 23000
430 FG=0:X=0:Y=CP/XC*NY:GOSUB 26000:X=NJ: GOSUB 26000:
    GOSUB 23000
440 FG=0:X=0:Y=CQ/XC*NY:GOSUB 26000:X=NJ: GOSUB 26000:
    A$=LEFT$(Q$,4):GOSUB 23000
450 FG=0:FOR X=0 TO NJ:Y=CO(X)/XC*NY:GOSUB 26000: NEXT X
460 IF NJ=0 THEN X=X+1: GOSUB 26000
470 A$="TOT": GOSUB 23000
480 GOSUB 29500: END
```

Figure 4.2   Listing for MIXING

balance is met by solving the balance equation to give the appropriate surface temperature. Because the equation is not linear, it has been solved using an iterative Newton's method which is described in Chapter 6. There are better and more sophisticated methods for estimating evapo-transpiration (see Chapter 3) and other individual components of the energy balance, but

```
Enter name of Source A: Through-flow
Enter conc of Through-flow: 800

Enter name of Source B: Overland flow
Enter conc of Overland flow: 100

Enter combined disch (Q) and Conc (C)
      or minus values to end input
```

0:	Q=100	: and C=600	QA=71	QB=29
1:	Q=140	: and C=500	QA=80	QB=60
2:	Q=200	: and C=350	QA=71	QB=129
3:	Q=280	: and C=200	QA=40	QB=240
4:	Q=350	: and C=250	QA=75	QB=275
5:	Q=370	: and C=300	QA=106	QB=264
6:	Q=360	: and C=400	QA=154	QB=206
7:	Q=325	: and C=500	QA=186	QB=139
8:	Q=285	: and C=600	QA=204	QB=81
9:	Q=235	: and C=650	QA=185	QB=50
10:	Q=205	: and C=750	QA=176	QB=29
11:	Q=−05			

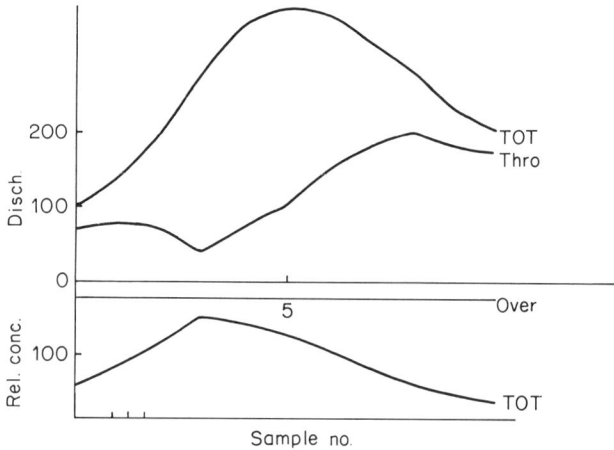

Figure 4.3   Example run for MIXING

RADBAL demonstrates the overall balance, and the system response to air and ground temperatures, albedo and surface wetness.

4.3   MASS OR ENERGY BALANCES INCLUDING ONE OR TWO STORES

In all cases where mass or energy can be physically held within a part of the system, then the storage equation (4.1) must include terms on the right-hand side of the equation for changes in storage amounts. In the simplest case there is a single store or reservoir. The STORFLO model described in Section 3.2 is an example of a simple reservoir model which is physically based and is capable of simulating a wide range of conditions. For STORFLO, the storage equation is given in equation (3.2), repeated here:

$$\frac{dS}{dt} = R - E - Q \tag{3.2}$$

in which the storage S represents the total reservoir of soil moisture in a catchment, with an inflow of rainfall R and outflows of runoff Q and evapotranspiration E.

4.3.1 A single-store catchment hydrology model

In STORFLO, the relationship between discharge and outflow is given by the exponential relationship of equation (3.1). A second example of a catchment hydrological model in which the total soil moisture is lumped into a single store is given by the listing for IEM4 in Figure 4.5. Equation (3.2) again applies, but the model differs in the introduction of a delay between rainfall and response; and in the dependence of discharge on moisture storage. For IEM4 this takes the form:

$$Q = \left(\frac{S}{A}\right)^2 \tag{4.4}$$

where discharge Q is measured in mm h^{-1}, storage S is in mm, and the constant A has the units of (mm h)$^{1/2}$. The program solves this version of the storage equation separately for the three cases of positive, zero and negative net rainfalls, the latter corresponding to rainfall at less than the given evaporation rate. The model also incorporates a regression equation which estimates the proportion of rainfall which contributes runoff, RP, in terms of the current soil moisture deficit, SD:

$$RP = PE \exp\left(-\frac{SD}{M}\right) \tag{4.5}$$

where PE is the proportion contributing when the catchment is, on average, at saturation; and M is a soil parameter with units of mm.

Table 4.1 gives optimized parameter values for a range of British catchments, from the Institute of Hydrology's *Flood Studies Report* (NERC, 1976), on which IEM4 is based. These values will allow you to investigate variations in the shape of observed flood hydrographs. The program offers values for the River Tyne, near Edinburgh, and includes the rainfall test data from the report for this catchment. Alternatively you may enter a rainfall sequence from the keyboard.

Single store models, although useful, are limited in scope. In the context of catchment hydrology, a single storage value must be used to represent all

```
 10 GOSUB 10000: REM Initialize
 20 RESTORE: READ SS ,AL,TA,TG,KE
 30 READ KG,KA
 40 HR=15:DIM RI(HR),T(HR),E(HR):RMAX=-1E9
 50 FOR H=0 TO HR
 60   READ A: IF A>RMAX THEN RMAX=A
 70   RI(H)=A: NEXT H
 80 FOR I= 1 TO 18: READ A$: A=(40-LEN(A$))/2:A$=LEFT$("
 90 GOSUB 19010
100 GOSUB 15000:R=0:C=0:C$="Albedo (%)":  UL=100: LL=0: A=AL:
    GOSUB 12500: AL=A:R=1
110 C$="Air Temp (Deg C)": LL=-10: UL=60: A=TA: GOSUB 12500:
    TA=A:R=2
120 C$="Ground Temp": A=TG:GOSUB 12500: TG=A:R=3
130 C$="% Potential E-T": LL=0: UL=100: A=KE: GOSUB 12500:
    KE=A
140 GOSUB 20000: GOSUB 560: REM Titles
150 NX=0:XX=17:NY=-15:XY=35: GOSUB 21000: SX=3: GOSUB 22000
160 X=7:Y=3:A$="Time (Hrs)":GOSUB 23000:X=0.1:Y=XY-4:
    A$="Temp":GOSUB 23000
170 FG=0:X=HR+.5:Y=35: GOSUB 26000 :X=-.5:GOSUB 26000
180 FOR H=0 TO HR:X=H-0.5:Y=35-RI(H)*30/RMAX:GOSUB 26000:
    X=X+1:GOSUB 26000:NEXT H:A$="Rad":GOSUB 23000
190 FG=0:T=TA:DT=0.01: ER=1E-4
200 EMAX=0:FOR H=0 TO HR
210   RR=RI(H)
215   REM Iterative solution using Newton's method
220   TX=T+DT/2:GOSUB 500:DF=FT:TX=T-DT/2: GOSUB 500:
      DF=(DF-FT)/DT
230   TX=T: GOSUB 500:XT=FT/DF: IF ABS(XT)>ER THEN T=T-XT:
      GOTO 220
235   R=2: C=0: A$="Temp at Time "+STR$(H)+" is "+
      STR$(INT(T+.5))+"  ": GOSUB 16000
240   X=H:Y=T:GOSUB 26000
250   EA=KA*KE/100/FNB(T)*1.436E-3
260   IF EA>EMAX THEN EMAX=EA
270   T(H)=T: E(H)=EA: NEXT H
280 A$="Temp":Y=Y+1.5:GOSUB 23000
290 FG=0: X=0:Y=TA:GOSUB 26000: Y=Y+1.5:A$="AIR":GOSUB 23000:
    FG=0:X=0:Y=TA:GOSUB 26000: X=15:GOSUB 26000
300 FG=0: X=0:Y=TG:GOSUB 26000:Y=Y-.3:A$="SOIL":GOSUB 23000:
    FG=0: X=0:Y=TG:GOSUB 26000: X=15:GOSUB 26000
310 IF EMAX=0 THEN GOTO 330
320 FG=0: FOR H=0 TO HR: X=H:Y=-E(H)*10/EMAX:GOSUB 26000:
    NEXT H:A$="Evap":Y=Y+1.5:GOSUB 23000
330 R=3:A$="":GOSUB 16000:GOSUB 29500:GOSUB 19010:
    GOSUB 15000: GOSUB 560
340 R=4: C=0: A$="HOUR   RAD'n      Temp(C)     Evap(mm/h)":
    GOSUB 16000:R=6
350 FOR H=0 TO HR
360   C=0:B=2: A=H: GOSUB 17500:B=3: D=1: A=RI(H): C=6:
     .GOSUB 17200
370   C=16: A=T(H): GOSUB 17200
380   C=26: D=3:A=E(H): GOSUB 17200: GOSUB 19000
390   NEXT H
400 GOSUB 29000:GOSUB 19010: GOTO 100
410 REM Data for Stefan-Boltzmann Radiation constant
420 DATA 5.67E-8,25,15,10,100
430 REM Data for ground and air loss rates
440 DATA 2,20
450 REM Data for hourly radiation values through day (W/SQ M)
460 DATA 0,10, 50,150,300,500,700,850,950,990,1000,990,980,
    960,840,720
```

(continued)

```
470 REM Titles
480 DATA "RADIATION BALANCE ROUND THE CLOCK"," ",
    " For given hourly radiation values"
485 DATA "Program gives surface temperature",
    "and Evapo-transpiration","depending on"
490 DATA "values you enter","for Albedo",
    "Air and Ground temperatures"
495 DATA "and how wet or dry the soil is",
    "in terms of percent Actual"
496 DATA "of Potential Evapo-transpiration"," ",
    "Try different values"
497 DATA "to see how much warmer","the ground gets",
    "if the soil is dry","or dark or hot"
500 REM Bowen's ratio is (TA-TX)*BB: FT is fn(TX) which
    should be zero for balance
510 IF TX=TA THEN BB=0.39: GOTO 530
520 BB=0.08+0.20*(1-EXP(1.526*(TA-TX)))/(TX-TA)
530 FT=RR*(AL/100-1)+SS*(TX+273)^4+(TX-TG)*KG+KA*
    (TX-TA+KE/100/BB)
540 RETURN
550 REM Titles subroutine
560 GOSUB 18000
570 PRINT"Albedo = ";STR$(AL);"%: Evap at ";STR$(KE);
    "% of Potential"
580 PRINT"Air and Soil Temps are ";STR$(TA);"C and ";
    STR$(TG);"C"
590 RETURN
```

Figure 4.4   Listing for RADBAL

```
 10 GOSUB 10000:REM INITIALIZE SCREEN
 20 FOR R=1 TO 7: READ A$:C=(40-LEN(A$))/2: GOSUB 16000:
    NEXT R: R=R+1:C=0
 30 C$="Basin Name: ":A$="R. Tyne Nr Edinburgh": GOSUB 11000:
    N$=A$
 40 R=R+1:GOSUB 580:AR=A
 50 GOSUB 580:PR=A/100
 60 GOSUB 580:M=A
 70 GOSUB 580:AC=SQR(A*M)
 80 GOSUB 580:DL=A
 90 GOSUB 580:QI=A
100 GOSUB 580:SD=A
110 GOSUB 580: EV=A/24
120 GOSUB 580:NT=A
130 REM ====================
140 REM Main calculation loop
150 ND=INT(DL):FD=DL-ND
160 DIM RA(NT),Q(NT)
170 C=0:A$="Enter 'T' for Test Rainfall":GOSUB 16000: R=R+1:
    C=6
180 C$="'H' to enter by hand: ":A$="T": GOSUB 11000:
    PRINTA$,LEN(A$):IF A$<>"T" AND A$<>"H" THEN GOTO 180
190 C$=A$:A$="Enter rainfalls (mm), ending with minus value"
200 RG=0: IF C$="T" THEN RG=2:
    A$="Rainfalls taken from FSR Test Data"
210 C=0: R=0:GOSUB 15000:RR=0: GOSUB 16000:XQ=0 :XR=0:
    Q(0)=QI
220 R=1:A$="Time(Hrs)    Rain(mm)         Disch(cumecs)":
    GOSUB 16000:R=2
```

(continued)

```
230 K=0: GOSUB 740: REM Print-out
240 QA=QI*3.6/AR:QB=1+FD*SQR(QA)/AC:QB=QA/QB/QB:
    CP=SD+AC*SQR(QA)
250 FOR K=1 TO ND-1
260   H=QB:QB=1+SQR(H)/AC:QB=H/QB/QB:QC=H*FD+QB*(1-FD)
270   Q(K)=QC*AR/3.6: GOSUB 740: NEXT K
280 FOR K=ND TO NT
290   RP=PR*EXP(-(CP-AC*SQR(QB))/M): IF RP>1 THEN RP=1
300   RE=(RA(K-ND)-EV)*RP
310   H=QB:IF RE=0 THEN QB=1+SQR(QB)/AC: QB=H/QB/QB: GOTO 350
320   IF RE<0 THEN E=SQR(-RE):QB=E*TAN(-E/AC+ATN(SQR(H)/E)):
      QB=QB*QB:GOTO 350
330   E=SQR(RE)/AC*2:E=EXP(E):Z=(E-1)/(E+1)
340   QB=RE*((SQR(QB)+(SQR(RE)*Z))/(SQR(RE)+(SQR(QB)*Z)))^2
350   IF QB<0 THEN QB=0
360   QC=H*FD+QB*(1-FD)
370   Q(K)=QC*AR/3.6:GOSUB 740
380   NEXT K
390 IF R>0 THEN GOSUB 29000:GOSUB 19010
400 REM ====================
410 REM Graph plot
420 GOSUB 20000: NX=0:XX=NT*1.3: NY=0:XY=XQ: GOSUB 21000
430 A$=N$: GOSUB 24000:X=XX/50:Y=XY*.65:A$="Disch":
    GOSUB 23000
440 GOSUB 22000
450 X=XX/2:Y=XY/10:A$="Time (Hours)":GOSUB 23000
460 FG=0:X=0: Y=XY: GOSUB 26000:X=NT:GOSUB 26000:
    FOR X=0 TO NT
470   RR=RA(X): IF RR=0 THEN GOTO 490
480   Y=XY*(1-RA(X)/XR/2):YA=XY:XA=X+1:GOSUB 27000
490   NEXT X
500 A$="Rain":GOSUB 23000
510 FG=0: FOR X=0 TO NT
520   Y=Q(X): GOSUB 26000: NEXT X
530 A$="Disch": GOSUB 23000: GOSUB 19010
540 GOSUB 29500:GOSUB 18000: FOR I=0 TO NT:RA(I)=0: NEXT I
550 A$="H": GOTO 190
560 REM ====================
570 REM Input subroutine
580 READ C$,A,LL,UL
590 GOSUB 12500: GOTO 19000
600 REM ====================
610 REM Data for titles
620 DATA Isolated Event Model,"",Forecasts hydrograph shape,
    and runoff sequence, from input rainfall sequence,"",
    Enter parameters as follows
630 DATA Basin Area (sq.km),307,1,10000
640 DATA % runoff at saturation,54,0,100
650 DATA Soil parameter (mm),220,1,1000
660 DATA Residence Time (hr),2,0.1,50
670 DATA Delay in hours,4.51,0,50
680 DATA Initial Q (cumecs), 2.3,0,1000
690 DATA Initial Deficit (mm),68,0,150
700 DATA Daily Evap (mm),1,0,5
710 DATA Total Duration (Hrs),60,5,500
720 REM ====================
730 REM Printout and rainfall input routine
740 C=6:A=K:B=3:GOSUB 17300
750 C=14:IF RG=0 THEN C$=">":A=RR:GOSUB 12000: RG=1:
    IF A>=0 THEN RG=0:RA(K)=A:RR=A : GOTO 770
760 IF RG=2 THEN READ RR: RA(K)=RR:A$=STR$(RR): GOSUB 16000
770 IF RR>XR THEN XR=RR
780 A=Q(K):IF A>XQ THEN XQ=A
790 C=30:B=2:D=2:GOSUB 17200: IF R=22 THEN GOSUB 29000
800 GOTO 19000
```

(continued)

```
810 REM =========================
820 REM Test Rainfall Data for R. Tyne from Flood Studies
    Report
830 DATA 0,0,0,0,0,0,0,0,.2,.5,.2,.5
840 DATA .5,2.8,.9,1.2,.2,.2,0,0,.5,0,0,.2
850 DATA 0,0,.3,0,0,0,.6,2.2,1.1,0,0,.3
860 DATA 2.8,.6,2.7,1,2.3,1.3,3,1.7,2,3,3.3,1.7
870 DATA 1.3,.7,0,0,0,0,0,0,0,0,0,0,0,0,0,0,0,0,0,0,0,0,0
```

Figure 4.5   Listing for IEM4

Table 4.1   Optimized parameter values for British catchments (adapted from *Flood Studies Report*, NERC, 1976)

FSR number	Catchment name	Optimum parameter values				
		Area (km²)	Runoff at sat'n (%)	Soil param. (mm)	Res. time (hours)	Delay (hours)
20/1	Tyne at E. Linton	307	54	220	1.86	4.51
24/3	Wear at Stanhope	172	53	1000	0.11	0.39
27/27	Wharfe at Ilkley	443	75	24	18.13	3.42
28/23	Wye at Ashford	154	72	25	265.62	2.37
37/1	Roding at Redbridge	303	100	72	67.62	3.52
39/17	Ray at Grendon U/wood	19	80	101	2.24	4.68
39/25	Enborne at Brimpton	153	59	50	15.31	5.27
40/10	Eden at Penshurst	224	78	71	14.61	7.31
41/5	Ouse at Goldbridge	182	78	95	9.01	8.31
45/2	Exe at Stoodleigh	422	94	75	31.16	2.33
46/3	Dart at Austins Br.	248	49	139	0.87	2.10
53/7	Frome at Tellisford	262	46	1000	0.30	5.41
54/16	Roden at Rodington	259	73	87	38.42	10.4
54/19	Avon at Stareton	347	63	120	10.03	21.8
54/22	Severn at Plynlimon Weir	8	69	585	0.28	0.09
57/4	Cynon at Abercynon	109	69	21	40.59	1.79
61/1	Western Cleddau at Prender-gast Mill	198	55	126	4.05	2.27
65/1	Glaslyn at Beddgelert	69	57	191	2.36	0.46
71/3	Croasdale Beck at Croasdale Flume	10	68	160	0.42	1.11
73/804	Brathay at Brathay Hall	57	89	305	2.10	4.47
76/11	Coal Burn at Coalburn	1.5	88	88	1.20	0.57

possible distributions of moisture over the area of the catchment and with depth through the soil profile. Improvements may be made in forecasting accuracy by increasing the number of stores. A catchment may be divided into two or more sub-catchments, each with its own parameters. The models for each sub-catchment are then run in parallel, and their outputs combined to give the total catchment discharge. Improved forecasting can also be achieved by representing the soil column by two or more stores, representing

```
 10 GOSUB 10000:REM Initialize
 20 V0=0: H0=0: FY=25
 30 CF=ATN(1)*2/3:YR=0
 40 DIM V(12),H(12):V(0)=V0:H(0)=H0
 50 C$="Uptake (kg/sq.m/yr)":A=2:LL=0.1:UL=10:GOSUB 12500:
    UR=A
 60 C$="% Leaf fall per year":A=10:LL=4:UL=100:GOSUB 12500:
    LR=A
 70 C$="% Decomposition /yr":A=20:LL=1:UL=60:GOSUB 12500:
    DR=A
 80 RR=120*UR/LR: RS=120*UR/DR: IF RS>RR THEN RR=RS
 90 GOSUB 20000:NX=-12:XX=12:NY=-RR/10:XY=RR
100 GOSUB 21000:SX=3:GOSUB 22000
110 X=-12:Y=0:A$="<<< Month <<<":GOSUB 23000:X=3:
    A$=">>> Month >>>":GOSUB 23000
120 Y=RR*.8:X=-12:A$="Veg Biomass":GOSUB 23000:X=1:
    A$="Organic Soil":GOSUB 23000
130 Y=RR:X=-2:A$="Kg/sq.m":GOSUB 23000
140 REM ============
150 REM Main loop
160 DIM VA(FY),HA(FY):FOR YR=0 TO FY
170    VA(YR)=V(0): HA(YR)=H(0)
180    FOR M=1 TO 12
190       GOSUB 430:GOSUB460:GOSUB 500:
          REM Uptake, leaf fall and Decomposition
200       V(M)=V(M-1)+UP-LF
210       H(M)=H(M-1)+LF-DC
220       NEXT M
230    REM =====================
240    REM Plot graphs
250    GOSUB 18000:PRINT "Year ";YR;:
       IF 2*INT(YR/2)<>YR THEN GOTO 280
260    FG=0:FOR M=0 TO 12: Y=V(M):X=-M:GOSUB 26000:NEXT M
270    FG=0:FOR X=0 TO 12:Y=H(X):GOSUB 26000:NEXT X
280    IF ABS(V(12)-V(0))<RR/1000 AND ABS(H(12)-H(0))<RR/1000
       THEN FY=YR: YR=50
290    V(0)=V(12):H(0)=H(12)
300    NEXT YR
310 GOSUB 29500:GOSUB 18000:GOSUB 19010
320 XX=FY*1.2:NX=0:NY=0:XY=110: GOSUB 22000: X=FY/30: Y=XY:
    A$="% Final Biomass": GOSUB 23000
330 X=FY/2:Y=10:A$="Years": GOSUB 23000
340 FG=0: FOR X=0 TO FY
350    Y=VA(X)*100/VA(FY): GOSUB 26000:IF X=INT(FY/3) THEN
       Y=Y+5:A$="Veg": GOSUB 23000:FG=0
360    NEXT X
370 FG=0: FOR X=0 TO FY
380    Y=HA(X)*100/HA(FY): GOSUB 26000: IF X=INT(FY/2) THEN
       Y=Y+5:A$="Org":GOSUB 23000: FG=0
390    NEXT X
400 GOSUB 29500: END
410 REM ================
420 REM Nutrient uptake rate in terms of month M(0-12)
430 UP=UR/12*(1-COS(M*CF)):RETURN
440 REM ============
450 REM Leaf fall rate in terms of month M and Veg biomass V
460 LF=LR/1200*V(M-1):IF M>8 AND M<11 THEN LF=LF+LF
470 RETURN
480 REM ============
490 REM Organic soil decomposition in terms of month M and
    Organic biomass H
500 DC=DR/1200*H(M-1)*(1-COS(M*CF)):RETURN
510 REM ============
```

Figure 4.6   Listing for GROWVEG

successive soil layers or simply unsaturated and saturated zones. In this case water is routed through the stores one after another: that is in series, the output of one store providing the input to the next.

4.3.2 A two-store nutrient cycling model

A biogeographical example of a two-store model is given in the listing for GROWVEG given in Figure 4.6. This model simulates the growth over the course of several years of the biomass in both the living vegetation and the organic soil. The currency of this model is kg m^{-2} of carbohydrate. The living vegetation store is increased by photosynthesis over an annual cycle. Its output is by 'leaf fall' (which includes loss of twigs, roots and stems), at a rate which is directly proportional to the living biomass month by month, but with a doubling of the rate in October and November to simulate the major period of autumn leaf fall. This leaf fall is directly input to the organic soil, which itself loses material by decomposition, at a rate proportional to the current amount of organic soil biomass. The model simulates development of the vegetation and organic soil, starting from a bare mineral soil.

The parameters offered by the program represent a temperate forest. The rate parameters may be changed to represent other major vegetation types. The rate of uptake depends on photosynthesis, increasing in hotter climates provided that water is plentiful. The rate of leaf fall depends mainly on plant size, rising to almost 100 per cent per year for herbs, and falling to 5 per cent for forests and perennial plants adjusted to arid or semi-arid conditions, all of which have a large ratio of trunk to leaves. For well-aerated soils, the rate of decomposition increases with temperature from about 20 per cent in temperate (10 °C) soils to 60 per cent for tropical (30 °C) soils, but dropping to almost zero if the soil is waterlogged, irrespective of temperature.

Another example of a two-store model in biogeography is the predator–prey model described in Section 3.5. In this case the Lotka–Volterra equations (3.20) and (3.21) are storage equations for the two stores, but their units differ, in being numbers of individuals of the predator and prey populations respectively. Thus the loss of prey through predation in equation (3.20) is given by the term c_1NP, whereas the corresponding gain in predator population is given by the term c_2NP in 3.22, and the ratio $c_1:c_2$ is the number of prey needed to support each predator.

4.4 LINEAR MULTI-STORE MODELS

Where the system to be modelled represents an area, then it is logical to break down the area into a series of stores, representing the state of successive points in the area. Where the area can be treated as a strip, such as a slope profile or a stream, then we require a chain of similar stores to describe the system, with flows leaving each store and entering the next. Where we need to describe a whole area, then we may need a two-dimensional network of stores. Such models are 'distributed' models, and

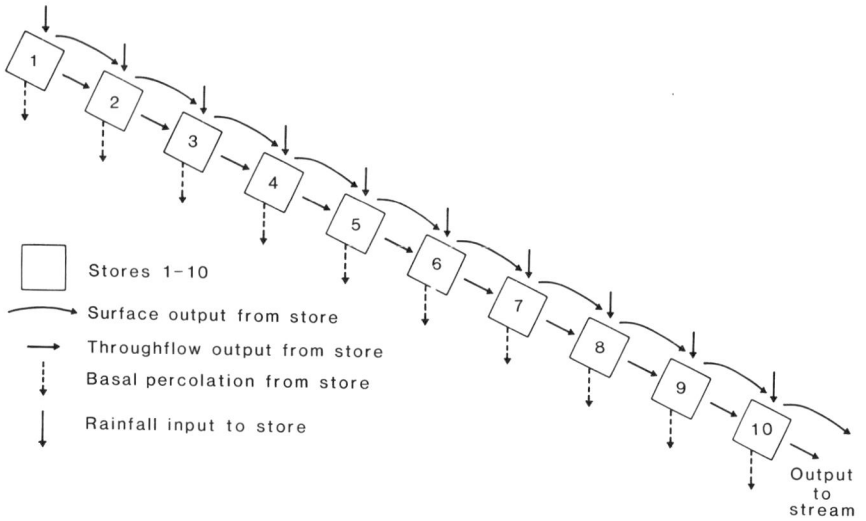

Figure 4.7 The hillslope system of ten linked soil profile stores for the model LINEAR

are generally more reliable than one- or two-store lumped models, although at the expense of greater complexity, both in the model and in the number of parameters needed.

4.4.1 A hillslope hydrology ten-store model

The model, LINEAR, represents ten stores linked together laterally to form a hillslope profile, which is described by the gradient and strip width at each point: increasing strip widths downslope represent divergent flow on a nose or spur, and decreasing strip widths represent a hollow. A diagrammatic representation of the stores in such a model is shown in Figure 4.7. Each store is capable of transmitting water downslope to the next store either as sub-surface throughflow or as overland flow, either because the storage capacity has been exceeded (saturation OF) or because the infiltration capacity has been exceeded (Horton OF). If a store becomes saturated it may influence the store upslope by preventing throughflow from entering it. In this way an area of saturation can build upslope, usually from the slope base.

The program listed in Figure 4.8 uses a series of sample files (CONCAVE, CONVEX, STRAIT and HOLLOW) for different slope configurations, each of which may be edited to modify gradient, strip width, storage capacity, initial percentage saturation, and upper and lower zone permeabilities for each of the ten store locations. A storage equation is worked out for each

```
 20 GOSUB 10000:REM initalise
 30 NH=60:DIM A(10),T(10),H(10),S(10),X(10),OF(NH),TF(NH),
    RF(NH)
 40 R=2: FOR I=1 TO 7: GOSUB 1000: NEXT I: REM Titles
 50 GOSUB 14550
 60 R=999:C=R: C$="Enter file name to load: ":A$="":
    GOSUB 11000:F$=A$
 70 DIM Z(5,10): GOSUB 14600
 80 GOSUB 18000: C=0:R=0:A$=" COL <1>    <2>   <3>
    <4>    <5>    <6>":GOSUB 16000
 90 R=1:A$="    Grad   Wdth   Cap    %Sat    K1    K2":
    GOSUB 16000
100 R=2::A$="   (Tan)   (m)    (mm)          m/day  m/yr":
    GOSUB 16000
110 R=3:A$= "Pt_____":
    GOSUB 16000
120 FOR I= 0 TO 10
130   R=I+4: C=0:A$=STR$(I): GOSUB 16000
140   FOR J=0 TO 5
150     C=6*J+4:GOSUB 14900: Z(J,I)=A:Q=A: GOSUB 940:
        GOSUB 16000
160     NEXT J:NEXT I : GOSUB 14700: FX=0
170 C=0: R=16:C$="'A' to accept: 'E' to edit: ":A$="":
    GOSUB 11000: IF A$<>"A" AND A$<>"E" THEN GOTO 170
180 IF A$="A" AND FX=0 THEN GOTO 360
190 IF A$="A" THEN GOTO 300
200 Q$=">": IF FX/2=INT(FX/2) THEN Q$=":"
210 FOR J=0 TO 5
220   C=0: R=17:C$="'C' accepts COL "+STR$(J+1)+": 'E'edits:
      ":A$="": GOSUB 11000: IF A$<>"C" AND A$<>"E" THEN GOTO 220
230   IF A$="C" THEN GOTO 280
240   C=J*6+3:C$=Q$: FOR I=0 TO 10
250     Q=Z(J,I): GOSUB 940
260     R=I+4:GOSUB 11000: Z(J,I)=VAL(A$)
270     NEXT I
280   NEXT J
290 FX=FX+1: GOTO 170
300 C=0: R=19: C$="'F' to save new file: 'C' to carry on: ":
    A$="":GOSUB 11000: IF A$<>"C" AND A$<>"F" THEN GOTO 300
310 IF A$="C" THEN GOTO 360
320 R=20: C$="Enter file name: ":A$=F$:GOSUB 11000:F$=A$:
    GOSUB 14500
330 FOR I=0 TO 10: FOR J=0 TO 5: A=Z(J,I): GOSUB 14800:
    NEXT J: NEXT I: GOSUB 14700
340 FX=0: GOTO 170
350 LL=20:UL=1000:A=100:C$="Slope Length (m)":GOSUB 12500:
    XL=A:DX=XL/10
360 C=0: R=21:LL=20:UL=1000:A=100:C$="Slope Length (m)":
    GOSUB 12500:XL=A:DX=XL/10
370 R=22:LL=0:UL=5:A=2:C$="Daily Evap (mm)":GOSUB 12500:
    EV=A/24:GOSUB 29000
380 REM ====================
390 REM Main program loop
400 HR=0:RR=0:GOSUB 15000: REM Clear text screen
410 MT=0: MO=0:XR=0:FOR I=0 TO 10: X(I)=Z(2,I)*Z(3,I)/100:
    NEXT I
420 C=0:R=0: A$="Enter Rf (mm): or - value to end":GOSUB
    16000
430 R=1: IF HR=0 THEN A$="Hour "+STR$(HR):GOSUB 16000:
    GOTO 470
440 C=0: R=1:A=RR:C$="Hour "+STR$(HR)+": RF = ":GOSUB
    12000:RR=A
```

(continued)

```
450 IF RR<0 THEN GOTO 730: REM to graph output
460 RF(HR)=A: IF A>XR THEN XR=A
470 IF HR=0 THEN R=3:C=0: A$="Pt  HorOF SatOF Store Tflo
    Perc  Qtot": GOSUB 16000
480 FOR J=0 TO 10
490   IF HR=0 THEN GOTO 590
500   JI=J-1:RJ=RR-EV
510   IF J>0 THEN W=Z(1,J)/Z(1,JI):T(JI)=T(JI)*W:S(JI)=S(JI)*W:
      H(JI)=H(JI)*W: X(J)=X(J)+T(JI): RJ=RJ+H(JI)+S(JI)
520   H(J)=0:KA=Z(4,J)/24: IF RJ>KA*1000 THEN H(J)=RJ-KA*1000:
      RJ=RJ-H(J)
530   X(J)=X(J)+RJ: IF X(J)<0 THEN X(J)=0: A(J)=0:T(J)=0:GOTO
      590
540   KB=Z(5,J)/24/365*1000:A(J)=KB:IF A(J)>X(J) THEN
      A(J)=X(J)
550   X(J)=X(J)-A(J)
560   S(J)=0:SC=Z(2,J):IF X(J)>SC THEN S(J)=X(J)-SC:
      X(J)=X(J)-S(J)
570   T(J)=X(J)*KA*Z(0,J)/DX: IF T(J)>X(J) THEN T(J)=X(J)
580   X(J)=X(J)-T(J)
590   R=J+5:C=0:A$=STR$(J):GOSUB 16000
600   C=4:Q=H(J):GOSUB 940: GOSUB 16000
610   C=10:Q=S(J):GOSUB 940:GOSUB 16000
620   C=16: Q=X(J):GOSUB 940:GOSUB 16000
630   C=22:Q=T(J): GOSUB 940: GOSUB 16000
640   C=28: Q=A(J):GOSUB 940: GOSUB 16000
650   C=34: Q=H(J)+S(J)+T(J): GOSUB 940: GOSUB 16000
660   NEXT J
670 C=0:R=17: IF HR=0 THEN A$="Total Overland flow at base
    = ": GOSUB 16000
680 B=3:D=3:C=30:A=H(10)+S(10):GOSUB 17200: OF(HR)=A : IF
    A>MO THEN MO=A
690 C=1: R=18: IF HR=0 THEN A$="Total Through-flow at base
    = ": GOSUB 16000
700 C=30:A=T(10): GOSUB 17200 :TF(HR)=A: IF A>MT THEN MT=A
710 HR=HR+1:IF HR<=NH THEN GOTO 440: REM for next hour
720 REM =========================
730 REM Graph output for OF and TF
740 HR=HR-1:GOSUB 20000: NX=0: XX=HR*1.2:NY=0: XY=MT*1.1:
    GOSUB 21000:GOSUB 22000
750 X=XX/50:: Y=XY: A$="TF Disch (mm/hr)": GOSUB 23000
760 X=XX/2: Y=XY/10: A$="Hours": GOSUB 23000
770 R=0: A$="Overland and Through-flow at slope base":
    GOSUB 24000
780 IF MO=0 THEN GOTO 810
790 FG=0: FOR X=0 TO HR:Y=OF(X)*MT/MO:GOSUB 26000: NEXT X
800 Y=Y+XY/20:Q=MT/MO: GOSUB 940: X=XX*.8:A$="OFx"+A$:
    GOSUB 23000
810 FG=0: FOR X=0 TO HR:Y=TF(X):GOSUB 26000: NEXT X
820 X=HR:Y=Y+XY/20:A$="TF": GOSUB 23000
830 FG=0:X=0: Y=MT: GOSUB 26000:X=HR:GOSUB 26000
840 Y=XY*.9:X=HR*.7: A$="RAIN":GOSUB 23000
850 FOR XA=1 TO HR
860   IF RF(XA)>0 THEN X=XA-1:YA=MT:Y=MT-RF(XA)/XR*MT/2:
      GOSUB 27000
870   NEXT XA
880 GOSUB 29500: C=0:R=2: A$="Try another rainfall sequence":
    GOSUB 24000: GOSUB 19010
890 FOR I=0 TO 10:A(I)=0:T(I)=0:H(I)=0:S(I)=0:NEXT I
900 FOR I=0 TO HR:OF(I)=0:TF(I)=0: RF(I)=0: NEXT I
910 GOTO 400
920 REM =======================
```

(continued)

```
930 REM Routine to put numbers into compact format
940 A$=STR$(INT(Q*1000+.5)):L=LEN(A$):IF L<4 THEN A$="."
    +RIGHT$("000"+A$,3): RETURN
950 IF L<6 THEN A$=LEFT$(A$,L-3)+"."+MID$(A$,L-2,6-L):RETURN
960 IF L<8 THEN A$=LEFT$(A$,L-3): RETURN
970 A$=LEFT$(A$,1)+"."+MID$(A$,2,1)+"E"+CHR$(L+45): RETURN
980 REM =======================
990 REM Routine to read and centre titles below
1000 READ A$:C=20-LEN(A$)/2:GOSUB 16000: R=R+1:RETURN
1010 REM =======================
1020 REM Data for titles
1030 DATA "10-STORE HILLSLOPE HYDROLOGY MODEL","","Forecasts
     overland flow"
1040 DATA "and two levels of sub-surface flow","down the
     length of a hillside"
1050 DATA "","Select a data file to use or edit"
```

Figure 4.8   Listing for LINEAR

of the ten soil moisture stores in turn, for each hour of the simulation. The input and output for each are:

INPUT

Rainfall (the same for all stores)

Horton overland flow, saturation overland flow and throughflow from upslope (except at divide: making allowance for changes in width between stores)

OUTPUT

Evaporation (the same for all stores if sufficient storage)

Horton overland flow (if net rainfall exceeds upper zone permeability)

Throughflow (at rate determined by upper zone permeability, gradient and depth of storage using Darcy's law)

Saturation overland flow (if storage exceeds capacity)

Percolation (at rate determined by lower zone permeability)

The storage equations are worked out in lines 510 to 580 of the program, with suitable checks to prevent the stores from going negative at any time. Figure 4.9 gives a sample data file (HOLLOW) and a resulting slope base hydrograph from a sequence of two simple storms. This multi-store model is more effective in forecasting the total response of a sub-catchment as well as giving detail about internal distributions of moisture. To simulate a whole catchment, however, a number of these hillside strips would need to run side by side. The model would also need to be extended to take account of water movement and storage in the channel, which is of considerable importance for catchments of more than about 10 km^2.

4.4.2   An *n*-store model for slope profile development

A second example of a linear multiple store model is given by the program SLOPEN, which is listed in Figure 4.10 and based on Kirkby (1971). It

COL Pt	⟨1⟩ Grad (Tan)	⟨2⟩ Wdth (m)	⟨3⟩ Cap (mm)	⟨4⟩ %Sat	⟨5⟩ K1 m/day	⟨6⟩ K2 m/yr
0	.280	10.0	100	.000	400	10.0
1	.260	10.0	100	3.00	400	10.0
2	.240	9.00	100	6.50	400	10.0
3	.220	8.00	100	11.0	400	10.0
4	.200	6.00	100	16.0	400	10.0
5	.180	4.50	100	22.0	400	10.0
6	.160	3.40	100	30.0	400	10.0
7	.140	2.60	100	38.0	400	10.0
8	.120	1.90	100	55.0	400	10.0
9	.100	1.40	100	67.0	400	10.0
10	.080	1.10	100	88.0	400	10.0

'A' to accept: 'E' to edit:A
'C' accepts COL 6: 'E' edits:C

'F' to save new file: 'C' to carry on:F
Enter file name: HOLLOW
Enter Slope Length (m) (20–1000): 100
Enter Daily Evap (mm) (0–5): 2

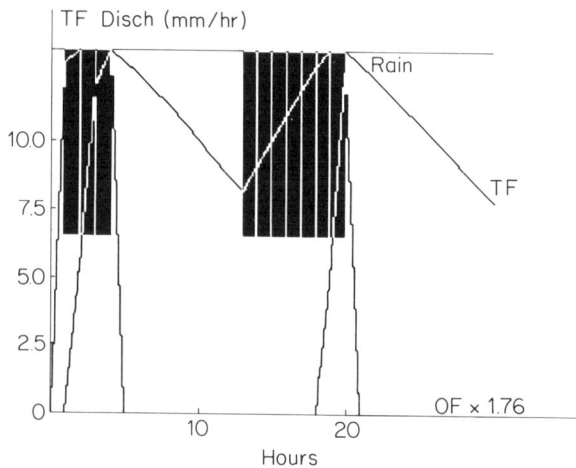

Figure 4.9 Example input and output for LINEAR (a) input data file for 'HOLLOW'; (b) slope base hydrograph from LINEAR slope model, using data from HOLLOW for a sequence of two simple storms

```
 10 GOSUB 10000:REM Initialize
 20 N=40:RESTORE:READ XT,YI,YN,GI,PM,K,U
 30 MN=N+N:DIM Z(MN),DZ(MN)
 40 GOSUB 15000:FOR R=0 TO 8: READ A$: C=20-LEN(A$)/2:
    GOSUB 16000: NEXT R
 50 T=0:UC=0:DC=0:GOSUB 590:REM Initial slope form
 60 GOSUB 20000:NX=0:XX=XT*1.3:RZ=XZ-NZ:NY=NZ-RZ/2:
    XY=XZ+RZ/4:GOSUB 21000:GOSUB 22000:REM Graphics screen
 70 X=XX/50:Y=XY:A$="Elevation (m)": GOSUB 23000
 80 Y=NY: IF NY<0 THEN Y=0
 90 X=XX/20: Y=Y+RZ/6:A$="Distance from divide (m)":
    GOSUB 23000
100 GOSUB 550 :CN=N:IT=0
105 :
110 GOSUB 18000: C=0:R=0:A$="Enter process rates":
    IF IT>0 THEN A$=A$+" or zero to finish"
120 GOSUB 16000:
130 R=1:LL=0:UL=500:A=K:C$="Creep etc (sq.cm/yr)":
    GOSUB 12500: K=A
135 IF K=0 THEN GOSUB 29000:END ELSE IF K<5 THEN GOTO 130
140 R=2:LL=10:UL=2000:A=U:C$="m. for Wash=Creep":
    GOSUB 12500:U=A
150 GOSUB 18000:C=0:R=0: A$="Stream Down- or Under-cutting
    (mm/1000yr)": GOSUB 16000
160 R=1:UL=1000:LL=-UL:C$="<1> DOWN ":A=DC: GOSUB 12500:DC=A
170 R=2:C$="<2> UNDER":A=UC:GOSUB 12500:UC=A
180 GOSUB 18000:C=0:R=0:LL=0:UL=1000:A=50:
    C$="1000's yrs to run":GOSUB 12500: TF=A*1000
190 R=1:UL=A:A=INT(A/10):C$="1000's yrs to plot":GOSUB 12500:
    TP=A*1000:NT=T+TP:TF=TF+T
200 GOSUB 18000: C=0:R=0: A$="TIME =            yrs":
    GOSUB 16000:R=1: A$="Iteration £        of": GOSUB 16000
210 :
220 REM Main loop for computing rates of change in storages
230 OZ=Z(0):FOR I=0 TO N
240   IF I=N THEN GOSUB 790:GOTO 290:REM Boundary condition
      at right to give DZ(N)
250   X=ABS(I+1/2-DV)*DX:NZ=Z(I+1):G=(OZ-NZ)/DX
260   GOSUB 850:REM Returns NS=Sediment transport from I as
      fn of G,X
270   IF I=0 THEN GOSUB 820:REM Boundary condition at left
      to give OS=Sediment into I=0
280   DZ(I)=(NS-OS)/DX:OS=NS:OZ=NZ
290   NEXT I:REM End of main loop
300 :
310 MAX=0:OZ=Z(0):OD=DZ(0):FOR I=1 TO N
320   NZ=Z(I):ND=DZ(I):H=ABS(OZ-NZ): IF H<1E-4 THEN GOTO 340
330   H=ABS(OD-ND)/H: IF H>MAX THEN MAX=H
340   OD=ND:OZ=NZ: NEXT I
350 DT=0.3/MAX: PF=0:IF T+DT>NT THEN DT=NT-T:PF=1
360 :
370 REM Routine to update elevations
380 T=T+DT:IT=IT+1:R=0:C=7:B=8:A=T: GOSUB 17300
390 R=1: C=12:B=4:A=IT:GOSUB 17300:C=22:A=DT:GOSUB 17500:
    C=999:R=C:A$=" Yrs   ":GOSUB 16000
400 XZ=-1E9:FOR I=0 TO N
410   Z(I)=Z(I)-DZ(I)*DT
420   IF Z(I)>XZ THEN XZ=Z(I): DV=I
430   NEXT I:IF UC=0 THEN GOTO 500
440 :
450 REM UNDER-cutting routine
460 A-INT(CN):CN=CN-UC/1E6/DX*DT: IF INT(CN)=A THEN GOTO 500
```

(continued)

```
470 A=INT(CN)+1:IF UC>0 THEN FOR J=A TO N:Z(J)=Z(N): NEXT J:
    GOTO 490
480 FOR J=N+1 TO A:Z(J)=Z(N): NEXT J
490 N=A
500 IF PF=1 THEN GOSUB 550: NT=NT+TP:IF NT>TF THEN NT=TF
510 IF T<TF THEN GOTO 230:REM Next Iteration
520 GOTO 110: REM New parameters for next time period
530 :
540 REM Plot profile
550 FG=0:X=0:FOR I=0 T N
560    Y=Z(I):GOSUB 26000:X=X+DX:NEXT I
570 RETURN
580 :
590 REM Sets up initial form of slope and max height, XZ
600 C=0:R=10:LL=0:UL=1000:A=XT:C$="Slope length (m)":
    GOSUB 12500:XT=A:DX=XT/N
610 R=11:UL=INT(XT/2+.50):IF YI>UL THEN YI=UL/2
620 A=YI:C$="Elev of divide":GOSUB 12500:YI=A
630 IF YN>UL THEN YN=0
640 R=12:C$="Elev of base":A=YN:GOSUB 12500:YN=A
650 R=13:UL=50:LL=-UL:A=GI: C$="Divide % Grad":GOSUB 12500:
    GI=A
660 R=14:LL=0:UL=100:A=PM:C$="% of length at "+STR$(GI)+"%":
    GOSUB 12500: PM=A
670 DV=0:NZ=1E9:XZ=-NZ:NM=INT(PM*N/100+.5)
680 FOR I=0 TO NM:Z(I)=YI-GI*I*DX/100:GOSUB 740:NEXT I
690 IF NM=N THEN RETURN
700 GM=(Z(NM)-YN)/(N-NM)
710 FOR I=NM+1 TO N:Z(I)=YN+(N-I)*GM:GOSUB 740:NEXT I
720 RETURN
730 REM Updates max & min  elevation
740 IF Z(I)>XZ THEN XZ=Z(I):DV=I
750 IF Z(I)<NZ THEN NZ=Z(I)
760 RETURN
770 :
780 REM Boundary condition at I=N
790 DZ(N)=DC/1E6:RETURN: REM DOWN-cutting
800 :
810 REM Boundary condition at I=0
820 OS=-NS:RETURN:REM Divide
830 :
840 REM Sediment transport out, in terms of distance X and
    gradient G
850 A=X/U:NS=K*1E-4*(1+A*A)*G:RETURN:REM Creep at rate K
    SQ CM/Y: Wash proportional to X*X, equal to wash at U m.
860 :
870 REM Data for Total slope length in m., Initial heights
    of left and right boundaries, % Gradient down from left,
    and % of slope at this gradient
880 DATA 200,100,0,1,90
890 REM Data for Creep rate (cm*cm/yr), Distance at which
    wash=creep (m)
900 DATA 10,100
910 REM Data for Titles
920 DATA *****************************,
    *   SLOPE EVOLUTION OVER TIME   *,
    *****************************,"",
    by Wash (Soil Erosion by water)
930 DATA and Creep/ Rainsplash/ Solifluction,"",
    Choose Initial form of slope profile,
    process rates and Slope Base conditions
940 REM +++++++++++++++++++++++++++++++++++++
```

Figure 4.10   Listing for SLOPEN

shows how a particular hillslope profile will develop over time under the action of specified rates of wash (soil erosion by water) and rainsplash or creep or solifluction. In this model, successive stores again represent successive sections of a hillslope profile, but in this case sediment movements are being budgeted, and the store of sediment is taken as the height of the hillside at each point. For any one out of a total of n stores, the mass balance equation may be written:

$$(S_1 - S_0) \, Dt = DzDx \qquad (4.6)$$

where S_1 = sediment input from upslope (except at divide)
         S_0 = sediment output to next store downslope
         Dt = time increment
         Dz = increase in storage (i.e. elevation)
         Dx = distance increment (i.e. distance between stores)

Sediment flow is computed from empirical process laws which express it in terms of gradient and distance from the divide (which is used as a measure of collecting area for overland flow on the slope). The same expression, with appropriate (and usually different) gradients and distances, is used to calculate both input and output of sediment. The only special cases are for the first (usually a divide) and the last (basal) stores. At the divide it is assumed that equal and opposite sediment flows leave in both directions; while at the base a river is assumed to carry away all the sediment delivered to it. The position of the basal stream is allowed to move laterally and or vertically at a constant rate. After a period of fixed conditions, the process rate parameters and/or the rate or direction of stream movement may be altered.

Process rates are calculated from the equation:

$$S = K \left[1 + \left(\frac{x}{u} \right)^2 \right] g \qquad (4.7)$$

where the sediment transport rate S is expressed in terms of distance x from the divide and local gradient g. K is a constant giving the rate of creep, splash or solifluction which give the first term in the equation, Kg, and u is the distance in metres beyond which the wash term, $K(x/u)^2 g$ becomes larger than the creep, etc. term. This equation is based on empirical measurements of soil creep, rainsplash and soil erosion rates, and the suggested constant values are thus based on direct process data, even if the form of equation (4.7) is somewhat simplified. The model generates an upper convexity in all cases, and a basal convexity or concavity according to the process balance and the rates and directions of river migration. Examples for a river moving laterally towards and away from the slope base are shown in Figure 4.11.

If too large a time step is used in calculating the erosion or deposition at any point, the model may 'overshoot', giving reverse slopes and becoming increasingly erratic in its behaviour. This kind of instability is entirely produced by using too large a time step in the model, and does not represent

Enter process rates or ZERO to end run
Enter creep etc. (sq. cm/yr) (0–500): 000

Enter process rates or ZERO to end run
Enter creep etc. (sq. cm/yr) (0–500): 000

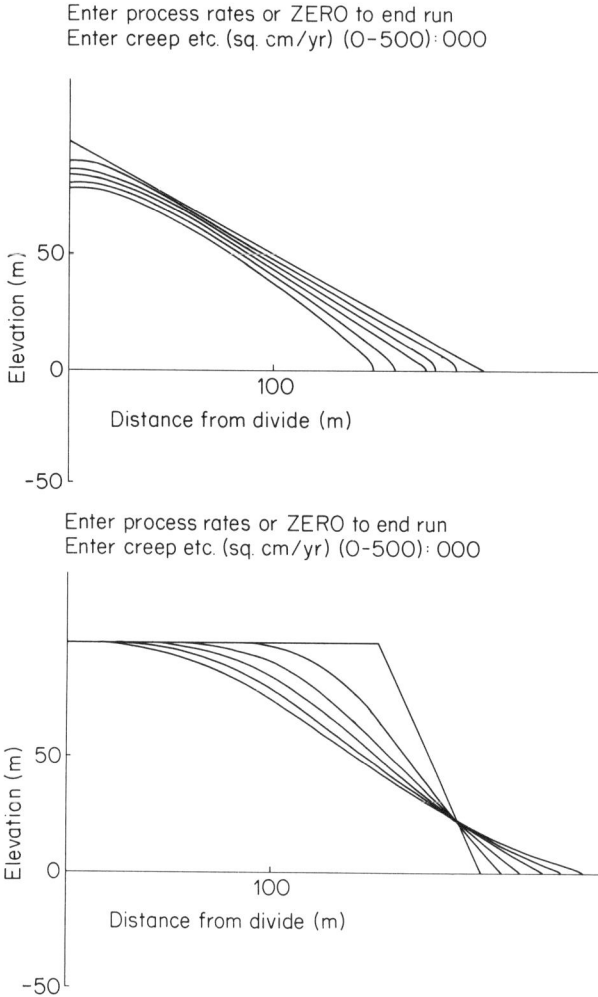

Figure 4.11   Example slope profile development sequences by creep/splash and soil erosion (a) river undercutting an initially steep slope; (b) river migrating away from slope base

a real response to the process rates over continuous time. To avoid this problem as parameter values are varied widely, the time step in SLOPEN is varied from step to step. The criterion used in the program, at lines 310 to 350, prevents the gradient changing by more than 30 per cent anywhere in any one step.

4.5   TWO-DIMENSIONAL NETWORKS OF STORES

The principles of linking stores together may be extended to a network of stores which represent sections of a two-dimensional surface. In general the flow of air, water, heat or sediment takes place in two, and often three dimensions. Although examples of such networks are beyond the scope of this book, exactly the same storage equation concepts may be applied to these more complex cases. Gradients, catchment areas and flows must all be defined in each direction, and care must be taken about the proper separation of downslope flows into their components.

There are some simpler networks where the flow directions are more constrained. For example flow from a catchment can be represented by a parallel set of similar models, each like the model, LINEAR, described above. By choosing flow strips following lines of greatest slope, there is no transverse flow from strip to strip, and flows from neighbouring strips only come together in the basal stream. Clearly, however, the growth of saturation up one strip has some effect on neighbouring strips. Boundary problems, such as this, increase the complexity of distributed models considerably. Another simplified network is the dendritic channel net. Each channel link has at most two tributary inflows together with channel precipitation and a hillslope inflow. Its outflows are down channel in a unique direction, together with evaporation and percolation losses. This is thus a much simpler system than a complete two-dimensional net.

The example programs in this chapter show how the storage equations for mass or energy can readily form the basis for a computer simulation model which forecasts the behaviour of a system over time. In all cases the quality of the forecasts depends strongly on the quality of the process data which must also be included in the models. The storage equation for mass or energy balance nevertheless provides a strong and indispensable framework, as well as a sound physical basis.

4.6 REFERENCES

Anderson, M. G. and Burt, T. P. (1978) The role of topography in controlling throughflow generation, *Earth Surface Processes*, **3**, 331–44.
Kirkby, M. J. (1971) Hillslope process response models based on the continuity equation, *Institute of British Geographers, Special Publication*, **3**, 15–30.
Lockwood, J. G. (1979) *Causes of Climate*. Edward Arnold, London.
Natural Environment Research Council (1976) *Flood Studies Report.*, Vol I: Hydrological Studies, HMSO, London.
Pilgrim, D. H., Huff, D. D. and Steele, T. D. (1978) A field evaluation of subsurface and surface runoff: II, Runoff processes, *Journal of Hydrology*, **38** (3/4), 319–41.

CHAPTER 5

Stochastic Models

5.1 WHY STOCHASTIC MODELS?

The models described so far are all deterministic, so that runs of a model using the same input and parameters always produce identical output. By introducing a stochastic or random element, successive runs no longer produce identical output but a distribution of outputs related to the stochastic components. The randomness may be introduced in the model input, in the processes or in parameter values, but in all cases the random element removes the unique relationship between input and output values.

To take a very simple example, let us imagine a system described by the linear relationship

$$Y = a + bX \tag{5.1}$$

where Y is 'explained' in terms of the variable X, for constants a and b, as described in Section 2.2. In practice this kind of relationship is derived from regression, and so has an uncertainty attached to it, described by a confidence band about the best estimate. Equation (5.1) should therefore be replaced by

$$Y = a + bX + r \tag{5.2}$$

where r is a random value or variable, with an average value of zero and a (normal) distribution around this value for which the spread is proportional to the width of the confidence band. Equation (5.2) is now a simple stochastic model. For each 'run' with the same input value of X, the outcome will be different according to the particular value of r. The values of Y obtained will be distributed in some way. In this simple case the distribution of Y will be very directly derived from the distribution of r, but where the relationships are not strictly linear, input and output distributions have a more complicated connection, and one which may not be apparent without using a simulation model.

In many cases, it seems appropriate to use randomly generated input for a model. Thus for the hydrological models described in Chapters 2 and 3, output hydrographs are generated from a sequence of rainfall input for successive time intervals. These rainfalls may be fed in as real data, as idealized data representing a design storm, or as a random sequence obeying certain rules. These rules might, for example, control the average, the spread and the degree of dependence on past values. The rules, or the values like averages which they depend on, may also vary through the seasons in a systematic way. This kind of stochastic model can be simply added into existing models as a change in the input routine, calling a stochastic subroutine instead of inputting values to the keyboard or reading them from data. Ways of generating random values and sequences are discussed in Section 5.2. Stochastic time sequences of this kind may be constructed to include linear trends and periodic variations. Their properties can thus be made to match the overall patterns obtained by analysing real data sets as described in Section 2.4, but with a different set of particular values each time the sequence is called up.

The input described above is generally a time sequence, and individual values are introduced into the simulation at successive time iterations of a model. Where models have a spatial dimension as well, as for many mass balance models, then it is often relevant to consider the effect of spatial variability in soil hydraulic or other properties. This may be done in three ways. First by running a deterministic model repeatedly, using slightly different values for one or more process parameters, drawn from a random distribution for each. This produces a spread of outcome forecasts, which gives the distribution of likely outcomes with a fairly small number of replicate runs, even if several parameters are stochastically assigned. A second approach is to run replicate models in parallel and combine their outcomes. Thus, for flow down a hillslope, different pathways through the soil may have different conductivities. By dividing the flow into any slope element into random or systematic components according to the frequency distribution of these conductivities, the individual flow velocities may be simulated, and recombined to give a total for the element. This approach is not exemplified here, but has considerable potential. A third method of incorporating spatial differences is applicable mainly to models with two horizontal dimensions of area. Distributed hydrological models, or models of landform evolution may fall into this category, and usually have a mass balance framework. Spatial distribution of soil and other properties may be incorporated directly through a stochastic distribution either of initial irregularities or of persistent process parameters. In both cases a spatial sequence must be generated, constrained by average values, spreads and dependence on neighbouring points. Appropriate stochastic models are to some extent a spatial analogue of the time sequences described in Section 5.2, but an alternative approach through fractals is outlined in Section 5.6.

The third, and perhaps most important kind of stochastic model is that in which the processes themselves have a stochastic component. Three examples are expanded below. In Section 5.3, a process of biological colonization is simulated for separated 'islands' of suitable habitat. The stochastic process consists of colonization of an area by a new species, which may or may not occur in any time interval. At each time step a random number is compared with the probability of colonization to determine whether it occurs or not. Any such process which describes only the probability of an event may be termed a stochastic process. Section 5.4 outlines a simplified model for gravel movement in streams, in which the probability of stone movement is controlled by whether the stone is free to move in various directions without obstruction by other stones or banks. Section 5.5 describes ways of simulating channel networks as a stochastic pattern, either in terms of draining a catchment area, or in terms of branching probabilities using the notion of a random walk, from which much of the mathematical work on stochastic processes has been developed.

The examples expanded below can only graze the surface of a large topic. Stochastic models tend to be more complex than deterministic models, commonly being built on top of them. Their development and application to physical geography has therefore been slower than for deterministic models, and all but simple examples are beyond the scope of this book.

5.2 GENERATING RANDOM VARIABLES AND SEQUENCES

Microcomputers usually contain a pseudo-random number generator, which generates values between zero and 1.0. Over this range, all values are equally likely to occur. The uniformity and randomness may be tested using the short program RANDOT (Figure 5.1(a)) which scales the screen to x and y coordinates in the range 0 to 1, and then plots dots with both coordinates chosen at random. The VDU screen should soon be covered with a roughly even scatter of dots, though the pattern is in detail random. You might note that the pattern is more irregular than most naturally occurring patterns because it lacks any tendency for neighbouring points to be alike.

If the listing is amended by replacing lines 210 to 230 with the versions in Figure 5.1(b) and (c), and the program is rerun, it will be seen that the dots are no longer evenly distributed, although still random. In practice, randomness does not imply that points are chosen haphazardly, but that they are drawn from a known frequency distribution. To convert uniformly distributed random numbers to random numbers from a target distribution, we must in general work back from the cumulative frequency of the target distribution, with its range of cumulative frequencies from zero to 1.0, to the desired random value. For example, an exponential distribution of mean x_0 has a cumulative frequency distribution:

$$P(x > x_1) = \exp\left(-\frac{x_1}{x_0}\right) \qquad (5.3)$$

The uniformly distributed random numbers give a series of equally likely cumulative frequency values within the range 0 to 1. The random value we want is x_1, which is derived from the uniform random number P as

$$x_1 = -x_0 \ln(P) \qquad (5.4)$$

```
(a) Listing for 'RANDOT'
  10 REM Random dot program
  20 GOSUB 10000: REM Initialize
  30 GOSUB 20000:REM Graphics Screen
  40 NX=0:XX=1:NY=0:XY=1: GOSUB 21000: REM Set Plotting ranges
  44 REM Draw Frame
  45 FG=0:X=0:Y=0:GOSUB 26000:YA=.999
  46 Y=YA:GOSUB 26000:X=YA:GOSUB 26000
  47 Y=0:GOSUB 26000:X=0:GOSUB 26000
  50 CO=1
  60 REM ====================
  70 REM Main loop
  80 C=0:R=0:A=CO:GOSUB 17000: REM Print Dot count
  90 GOSUB 200:X=RA:REM Random value
 100 GOSUB 200:Y=RA
 110 GOSUB 25000:REM Plot a dot @ x,y
 120 CO=CO+1:GOTO 80
 130 REM ==================
 140 END
 200 REM Random number generator
 210 GOSUB 14000
 220 RA=A
 230 RETURN
 240 REM =================

(b) Subroutine for exponential distribution
 200 REM Random numbers from exponential distribution with
     mean XM
 210 GOSUB 14000
 220 GOSUB 14100: REM Natural log
 230 RA=-XM*A: RETURN
 240 REM =================

(c) Subroutine for normal distribution
 200 REM Normally distributed random numbers: mean XM ,
     standard deviation SD
 210 RA=0: FOR I=1 TO 48:GOSUB 14000:RA=RA+A:NEXT I
 220 RA=(RA-24)/2*SD+XM
 230 RETURN
 240 REM =================

(d) Subroutine for Gamma(2) distribution
 200 REM Gamma-2 Distribution with peak at XM, mean at 2XM
 210 GOSUB 14000:RA=XM*SQR(1-A):B=A
 220 A=B/(1+RA/XM):GOSUB 14100: REM Natural Log
 230 A=-XM*A: IF ABS(A-RA)>1E-2 THEN RA=A: GOTO 220
 235 RETURN
 240 REM =================

(e) Subroutine for log-normal distribution
 200 REM Log-normal distributed random numbers: mean XM,
     standard dev. SD
 210 RA=0: FOR I=1 TO 48:GOSUB 14000:RA=RA+A:NEXT I
 220 RA=XM*EXP((RA-24)/2*SD/XM)
 230 RETURN
 240 REM =================
```

Figure 5.1 Listing for 'RANDOT' (a) complete listing including subroutine at lines 200 to 240 for uniformly distributed random numbers; (b) replacement subroutine for exponentially distributed random numbers; (c) replacement subroutine for normally distributed random numbers; (d) replacement subroutine for gamma(2) distribution; (e) replacement subroutine for log-normal distribution

This process is reflected in the listing of Figure 5.1(b). This inversion from the cumulative distribution is not always easy to perform. Figure 5.1(d) shows a rather slow iterative method for generating random numbers from the gamma(2) distribution, peaking at x_0 (with mean $2x_0$):

$$P(x>x_1) = \left(1 + \frac{x_1}{x_0}\right) \exp\left(-\frac{x_1}{x_0}\right) \qquad (5.5)$$

A different method can be used for normally distributed random numbers. It makes use of the central limit theorem which states that the sum of over 30 random values is always normally distributed, whatever the original distribution of the values used. Thus a sum of n uniformly distributed random values lying between 0 and 1 is normally distributed with mean $n/2$ and standard deviation $(n/12)^{1/2}$. This approach has been used in the listings of Figure 5.1(c) and (e) to give random values with normal and log-normal distributions respectively. For example, if the program to calculate grain movement thresholds on a stream bed (Figures 3.9 and 3.10) were used as part of a grain movement simulation, the routine in Figure 5.1(c) would give normally distributed random values. With the appropriate mean (XB) and standard deviation (SD), we obtain a random velocity from the distribution. If this velocity exceeds the threshold value, movement begins for the grain in question. Repetition of such random values provides one way (although not an efficient one) of assessing the probability of motion.

The methods described above give independent random values drawn from a chosen distribution. In some cases these values are all that is needed as the random input for a model, but in many cases random inputs need to form a sequence with some mutual dependence between successive values. Some of the simple sequences which may be produced are demonstrated in the program RANSEQ, listed in Figure 5.2. In its listed form, it creates a sequence of 100 independent values, with no mutual dependence at all. In this case the individual values are uniformly distributed over the range -1 to $+1$, although most of the distributions discussed above are also suitable. By changing the constant PP in line 25 to take a larger value (between zero and 1.0), the sequence becomes a simple or first-order Markov chain, in which successive values are related by

$$y_{n+1} = PP\, y_n + R\,(1-PP) \qquad (5.6)$$

where $y_0, y_1, \ldots, y_n, y_{n+1}, \ldots$ are successive terms and R is an independent random value. As PP rises from zero (complete independence) to 1.0 (complete dependence), the sequence shows more gradual variations, and longer runs of successive high or low values. If the sequence of random values at $0,1,2, \ldots$ steps ago is given by $R_0, R_1, R_2, \ldots$, the current term in the sequence is

$$R_0(1-PP) + R_1(1-PP)^2 + R_2(1-PP)^3 + \ldots$$

```
 10 GOSUB 10000:REM Initialize
 20 GOSUB 20000:REM Graphics screen
 25 N=100:PP=0
 30 NX=0:XX=N:XY=1:NY=-XY
 40 SX=N/5:SY=1:GOSUB 22000:REM Scale axes
 50 DIM Q(N)
 60 C=0:R=0:B=4:FOR X=0 TO N
 70   GOSUB 200: REM Independent random values with zero mean
 80   Q(X)=RA
 90   A=X:GOSUB 17300::REM Print count
100   NEXT X
110 FG=0:X=0:Y=0:GOSUB 26000
120 FOR X=1 TO N-1
130   GOSUB 300:REM Create sequence
140   GOSUB 26000:REM Plot sequence
150   NEXT X
180 C$="Press RETURN key to repeat":A$="":GOSUB 11000:
    GOSUB 18000:GOTO 60
190 REM =================
200 REM Random numbers with zero mean
210 GOSUB 14000:RA=A*2-1
220 RETURN
230 REM =================
300 REM Convert Independent values to a sequence
310 Y=PP*Q(X)+(1-PP)/2*(Q(X-1)+Q(X+1))
320 RETURN
350 REM =================
```

Figure 5.2   Listing for RANSEQ

so that previous random values have a decreasing influence over time. Longer persistence of influence may be obtained by generalizing equation (5.6) to include y_{n-1}, etc. terms on the right-hand side (second- or higher-order Markov chains).

A second process for producing a smoother series of input values is by applying some kind of weighted moving average to the original sequence of random numbers. If line 310 in the RANSEQ program is replaced by

$$310 \; Y=PP*Q(X)+(1-PP)/2*(Q(X-1)+Q(X+1))$$

then various values of PP from 0 to 1 produce different degrees of smoothing, with the greatest effect at PP$=\frac{1}{3}$. This arithmetic moving average (ARIMA) process can provide greater smoothing, either by repeating the smoothing operation on the sequence two or more times, or by operating on a wider range of the original random values. In the example here, the operation is only on the immediate value and one either side of it, but this range can be extended.

A third way of generating a random sequence of values is as a random walk. In this case the persistence is very much greater, in fact technically infinite, and the resulting sequence is not very well behaved statistically, with poorly defined mean and variance. These aspects of its behaviour arise because there is no tendency for the walk to return to its original starting point, the best estimate of future position always being the current position!

It is nevertheless easily generated and appropriate in some cases. To generate a random walk on the screen, change the value of XY in line 30 from 1 to 'SQR(N)', and change line 310 to

$$310\ Y = Y + Q(X)$$

The greater persistence of this series can be seen more clearly if the value of N in line 25 is increased from 100 to at least 500 (though not more than about 2000). In fact the random walk is an easily generated extreme example of a class of fractal sequences all of which have indefinite persistence. This topic is pursued in Section 5.6 in the context of a two-dimensional example.

This section has described some of the techniques used to generate random values and sequences, and the programs described include simple algorithms which may be incorporated in many of the deterministic models of previous chapters. These methods will also be used in the geographical examples described next.

5.3 ISLAND COLONIZATION

A chain of volcanic islands is progressively colonized from the nearest mainland source of biological material. Similar concepts are involved in the colonization of recently deglaciated areas from unglaciated refugia; or the spread of humid species back into formerly arid areas following climatic reversals. The spread of infection between communities is another similar process. In all cases, there are a finite number of separated areas, each described by its area, position and the number of source species currently found on it. In the case of a continuous area, it may be subdivided into convenient sub-units. Normally such areas are distributed across a map in two dimensions, but a one-dimensional model is used here to illustrate the concepts involved.

Fresh colonization depends on a number of factors, which are well discussed in a theoretical context by MacArthur and Wilson (1967). It is clearly a stochastic process, depending on chance factors of currents, wind and faunal movements, but it is also clear that a number of known factors influence the likelihood or rate of colonization. The first factor is the size of the potential destination area, and it is rational to suppose an increase with area, perhaps with saturation towards some upper limit; linear proportionality might be a first approximation. A second significant factor influencing rate of colonization is the distance between source and destination. Other factors being equal, then greater distance should lessen colonization rate. This effect expresses both the increasing perimeter of a circle as its radius increases if potential colonists spread out in all directions, and the expected reduction in number of colonists able to effectively disperse and survive over increasing distance. Thus, for example, swimming colonists are unlikely to have the range of wind dispersed species. Distance effects may also be influenced by the presence of intermediate islands, although such effects are more difficult to simulate, and will be ignored here. An inverse square law is used here as an exploratory hypothesis, representing the increasing perimeter factor on its own. The third important control on colonization is the number of species in the source and destination areas. This control acts directly through the potential number of new migrant species; and also by changing the probability that any incoming species will locate a suitable unoccupied niche. The first of these effects should give a rate of colonization directly proportional to the number of species in the source area which are not found in the destination area, and may be approximated (as here) by the difference in the respective species numbers. The niche availability factor is likely to decrease as number of species in the target area increases. For exploratory purposes this rate has been taken as inversely proportional to $(1 + N_D)$, where N_D is the number of species in the destination area, and the value of 1 has been added to provide finite values when $N_D = 0$.

Combining these effects, the proposed model for the average rate or expected number E of new species in unit time is given by:

$$E \propto A_D \frac{N_S - N_D}{(1 + N_D) \, X^2} \qquad (5.7)$$

where A_D is the area of the destination island, N_S, N_D are respectively the source and destination species numbers, and X is the distance between source and destination islands. Expected numbers of colonizer species may be obtained for each possible combination of source and destination. If expected numbers are taken as actual numbers, then the model described is strictly deterministic. The stochastic element lies in converting expected to actual numbers. Since the process is thought to be constant in rate over time, with colonization occurring with low frequency, then the numbers of colonist species should follow a Poisson distribution:

$$p(n) = \left[\frac{E^n}{n!}\right] \exp(-E) \qquad (5.8)$$

where $p(n)$ is the probability of exactly n new species. Each migration path should be treated as a distinct Poisson process, so that a random number should be used to convert expectations (from equation (5.7) above) to actual numbers colonizing from each source to each destination. These stochastic values should then be combined to give simulated net migration rates.

Figure 5.3 lists a program, COLONY, which carries out this simulation for a chain of nine islands, loosely representing the Canaries out to the Azores. All the islands begin the simulation with no mainland species. Line 500 expresses the process function of equation (5.7) above. Lines 3700 to 410 implement the Poisson distribution (equation (5.8)) working from a random number A between 0 and 1. Figure 5.4 shows the initial values used and the distribution at time 20. It may be seen that there is a species gradient away from the source area. You may like to vary the process function in line 500, for example lessening the fall-off with distance to account for shipboard transport; and the size and spacing data beginning from line 550. Another appropriate example for study might be based on the Caribbean Islands from Florida out towards the Leeward Islands.

This model may be developed in at least two directions. First, it might use *x* and *y* coordinates to define position in two dimensions rather than a single spatial dimension. Second a species list could be included for each island (in the form of a string variable), so that the number of possible

```
 10 REM Island Colonization program: COLONY
 20 GOSUB 10000:REM Initialize
 30 C=6:R=0:A$="Island Colonization":GOSUB 16000:C=0:R=2:
    A$="ISLAND        AREA   DIST SPECIES   NEW":GOSUB 16000
 40 C=0:R=22:A$="Time 0":GOSUB 16000
 50 RESTORE:READ NC:DIM Q$(NC),N(NC),DN(NC),D(NC),AR(NC)
 60 READ N(0),TP,K:D(0)=0
 70 T=0
 80 B$="                           "
 90 FOR I=1 TO NC :REM Read initial values & tabulate to
    screen
100    READ  Q$(I),AR(I),D(I),N(I)
110    R=I+3:C=0:A$=Q$(I):GOSUB 16000
120    C=14:A=AR(I):B=5:GOSUB 17300
130    C=20:A=D(I):GOSUB 17300
140    GOSUB 420
150    NEXT I
160 REM ====================
170 REM Main loop
180 IF INT(T/TP)=T/TP THEN GOSUB 29000:
    REM Copy screen to printer
190 T=T+1:C=5:R=22:A=T:B=4:GOSUB 17300
200 C=5:R=23:A$=B$:GOSUB 16000
210 FOR J=0TO NC:NS=N(J):XS=D(J)
220    FOR I=1 TO NC
230       IF I=J THEN GOTO 270
240       EX=FNP(ABS(D(I)-XS)):GOSUB 14000:REM Random No
250       GOSUB 380:REM Convert expectation to actual moves
260       IF NQ>0 THEN DN(I)=DN(I)+NQ:GOSUB 330:
          REM New Colonists
270       NEXT I
280    NEXT J
290 FOR I=1 TO NC:GOSUB 420:REM Update numbers
300    NEXT I
310 GOTO 180
320 REM ===================
330 REM Report Moves as they occur
340 C=0:R=23:A$="Move "+STR$(NQ)+" from " + STR$(J) +
    " to " + STR$(I) +"       ":GOSUB 16000
350 RETURN
360 REM ======================
370 REM Poisson distribution
380 B=EXP(-EX):Q=B :U=B::NQ=0
390 IF A<Q THEN RETURN
400 NQ=NQ+1:U=U*EX/NQ:Q=Q+U:GOTO 390
410 REM Should always exit at 400
420 REM ===================
430 REM Update Number of Species
```

(continued)

```
440 N(I)=N(I)+DN(I)
450 R=I+3:B=4:IF T>0 THEN C=34:A=DN(I):DN(I)=0:GOSUB 17300
460 C=27:A=N(I):GOSUB 17300
470 RETURN
480 REM ====================
490 REM x=Dist from source to destination:
    NS=No of species at Source
500 DEF FNP(X)=AR(I)*(NS-N(I))/(1+N(I))/X/X*K
510 REM =====================
520 REM No of colonies; No of species in Source area;
    Print frequency
530 REM and Scaling constant for probability in line 510
540 DATA 9,1000,50,0.4
550 REM Name; Area; Distance from Source;
    Initial Number of species;for each colony in turn
560 DATA Fuertaventura,1200,100,0
570 DATA Gran Canaria,1500,200,0
580 DATA Tenerife,2000,300,0
590 DATA La Palma, 500,400,0
600 DATA Madeira,600,700,0
610 DATA Santa Maria,100,1500,0
620 DATA Sao Miguel,500,1600,0
630 DATA Terceira,300, 1800,0
640 DATA Flores,100,2100,0
```

Figure 5.3   Listing for COLONY

colonizing species could be accurately determined. At a much more general level, this model demonstrates the use of a stochastic process to modify a model which could also be run deterministically. The value of a stochastic approach must be treated case by case. For the model described here, the greatest benefit of the stochastic approach appears to lie in the possibility of generating distributions of species numbers over replicate model runs. As you will quickly discover by experiment, these distributions show their greatest relative spreads when the actual numbers are low: that is in the early stages of colonization of each island.

5.4   KINEMATIC WAVES FOR GRAVEL BED-LOAD

The concept of kinematic waves as a model for pool and riffle spacing in gravel bed rivers has been introduced into the literature by Langbein and Leopold (1968). They argued that the stable position of riffles indicated a state of maximum gravel movement, and therefore an efficient organization of the channel to cope with its gravel load. They also explored a computer and laboratory simulation in which particles moved along a channel of one particle width. Movement of a grain was prevented if the space in front of it was occupied, and occurred with constant probability if unoccupied. Movement continued until prevented by another particle. In developments of this model, particles have a constant probability of stopping, even without impediment, as otherwise the entire sediment supply gradually builds into a single gravel bar for the channel section modelled.

102

```
(a)                    Island Colonization

      ISLAND              AREA @ km SPECIES EXTRAS

      Fuertaventura    1200    100.     0.
      Gran Canaria     1500    200.     0.
      Tenerife         2000    300.     0.
      La Palma          500    400.     0.
      Madeira           600    700.     0.
      Santa Maria       100   1500.     0.
      Sao Miguel        500   1600.     0.
      Terceira          300   1800.     0.
      Flores            100   2100.     0.

      Time 0


(b)                    Island Colonization

      ISLAND              AREA @ km SPECIES EXTRAS

      Fuertaventura    1200    100.    58.     1.
      Gran Canaria     1500    200.    24.     1.
      Tenerife         2000    300.    19.     0.
      La Palma          500    400.    11.     2.
      Madeira           600    700.     7.     1.
      Santa Maria       100   1500.     0.     0.
      Sao Miguel        500   1600.     2.     0.
      Terceira          300   1800.     0.     0.
      Flores            100   2100.     0.     0.

      Time     20.
      Move 1 from 0 to 5
```

Figure 5.4 Sample output from COLONY using parameter values assumed in the listing (a) initial conditions; (b) after 20 iterations

The model presented here develops these ideas to allow channels of width greater than a single grain. At each time interval, movement may occur over a range of downstream directions. As with the one grain channel, no movement occurs if the grain's path is obstructed in its chosen direction. Additionally, no movement occurs if the particle wishes to move into either bank. As for the one grain channel, downstream movement continues (now in randomly chosen directions) until obstructed or stopping by chance.

The exact course of the simulation is determined by a number of parameters, notably channel width (in grain units), the initial density of grains in the channel and probabilities for unobstructed particles to start or, once started, to stop. The probabilities of moving in various downstream directions are also important. In the program listed in Figure 5.5 (KIN-WAVE), movement is equally likely in three directions: directly downstream and downstream with one step to left or right. If the directly downstream direction is more likely, with probability PD (between $\frac{1}{3}$ and 1), and the diagonal directions each $(1-PD)/2$, then line 170 should be replaced by:

170 GOSUB 14000: A=INT((A+(PD−1)/2)/PD)

Alternatively, it may be more realistic to make these probabilities partly dependent on the distribution of stones, in that flow will tend to be greatest in less obstructed directions. This possibility has not been pursued, but is partly dealt with by the prevention of stone movement in obstructed directions.

Another factor to consider in setting up a simulation is the behaviour of the channel at its ends. In the listing, line 190, grains from the right-hand end of the channel section (NX=79) are re-introduced at the left-hand end (NX=0), providing the conditions of a simple recirculating flume. There is some danger in this simple approach, in that regularities may develop which are related to the length of the modelled section. An alternative which partially decouples the two ends is to count the number of stones leaving the right-hand end, and estimate the mean rate of sediment removal. This rate is then used as the expected rate of sediment input, using a Poisson distribution as in the colonization model above to give the actual number at each time step.

Figure 5.6 illustrates a run of KINWAVE with the parameters suggested in the listing. An initial random distribution (Figure 5.6(a)) of 20 per cent density, is changed into a more regular pattern with some evidence of clumping into five to seven bars (Figure 5.6(b)). By running the model, some evidence can be obtained about dependence of modelled bar spacing on channel width and probability parameters, and thence on whether the model represents at least some of the significant effects of gravel movement. To take the analysis further, a running average or similar smoothing procedure is needed to identify 'bars' in an objective way.

It is clear that a number of very important aspects of channel behaviour have been left out of this highly simplified model. For example, no allowance has been made for a range of grain sizes, and the model needs to be totally redesigned to include this factor. A second major omission is the possibility of building bedforms more than one grain deep, with consequences for flow direction and bed gradients: this is only one aspect of a complete neglect of the channel hydraulics. In some cases such highly simplified models are still able to tell us something about real world performance, usually in the cases where the process modelled is one of the most significant influences on the forecast outcomes.

In general terms, the KINWAVE model differs appreciably from COLONY in that KINWAVE could only be run meaningfully as a stochastic model, with no useful deterministic counterpart. Nevertheless, it does not produce

```
  10 GOSUB 10000:S$=CHR$(48)
  20 T=0:SU=0:HO=0:C=0:RESTORE:READ K,P,Q,W
  25 PRINT "2-D Kinematic wave for gravel bars"
  30 C$="Enter concentration (<1): ":A=K:R=3:GOSUB 900:K=A
  40 R=6:A$="Enter probabilities":GOSUB 16000
  50 R=7:C$="of moving: ":A=P:GOSUB 900:P=A
  60 R=8:C$="and of stopping again: ":A=Q:GOSUB 900:Q=A
  70 R=12:C$="Enter width of flow (integer): ":A=W
  80 GOSUB 12000:A=INT(A+.5):IF A<1 OR A>10 THEN GOTO 80
  90 W=A-1:DIM Z(79,W)
 100 GOSUB 500:REM Set up initial random distribution
 110 REM ========================
 120 REM Main loop
 130 GOSUB 14000:X=INT(80*A):GOSUB 14000:Y=INT(A*(W+1)):
     U=Z(X,Y)
 140 IF U=0 THEN GOTO 130:REM No Stone
 145 T=T+1:GOSUB 14000:IF A>P THEN GOTO 130:REM No movement
 150 OX=X:OY=Y:T=T+1
 155 REM ========================
 160 REM Inner loop for length of hops
 170 GOSUB 14000:A=INT(3*A)-1:REM Random value -1,0 or +1
 180 NY=Y+A:IF NY>W OR NY<0 THEN GOTO 220:REM Stopping against
     bank
 190 NX=X+1:IF NX=80 THEN NX=0:REM Recirculating 'flume'
 200 NU=Z(NX,NY):GOSUB 14000:IF NU=1 OR A<Q THEN GOTO 220:
     REM Stopping
 210 GOSUB 700:X=NX:Y=NY:U=NU:GOTO 170:REM Keep moving
 215 REM ========================
 220 H=X-OX:IF H<0 THEN H=H+80
 225 IF H=0 THEN GOTO 280
 230 REM Update after movement
 240 SU=SU+H:HO=HO+1:Z(OX,OY)=0:Z(X,Y)=1
 270 H=HO/10:IF H=INT(H) THEN GOSUB 800
 280 IF T/1000 = INT(T/1000) THEN GOSUB 800:GOSUB 29000:
     REM Screen dump to printer
 290 GOTO 130:REM End of main loop
 500 REM ========================
 510 REM Random initial pattern in Z array and on screen
 520 GOSUB 15000:PRINT "KINEMATIC WAVE FOR GRAVEL BARS"
 522 A$="==========":A$=A$+A$+A$+A$
 524 C=0:R=3:GOSUB 16000:R=14:GOSUB 16000
 526 IF W<9 THEN R=5+W:GOSUB 16000:R=R+11:GOSUB 16000
 528 R=2:A$="Flow direction >>>>>>>>>>>>>>>>>>>>>>":GOSUB 16000
 530 FOR TY=0 TO W
 540   FOR TX=0 TO 79
 550     GOSUB 14000: IF A<K THEN A$=S$:Z(TX,TY)=1:GOSUB 850
          GOSUB 16000
 560     NEXT TX
 570   NEXT TY
 580 GOSUB 29000:REM Screen to printer
 590 RETURN
 600 REM ========================
 700 REM Show movement on screen
 710 TX=X:TY=Y:GOSUB 850:A$=" ":GOSUB 16000
 720 TX=NX:TY=NY:GOSUB 850:A$=S$:GOSUB 16000
 730 RETURN
 740 REM ========================
 790 REM Update summary statistics
 800 C=0:R=0:A$="Time =            Transport =        ":
     GOSUB 16000
```

(continued)

```
 805 C=7:B=6:A=T:GOSUB 17300:C=28:A=SU:GOSUB 17300
 810 R=1:C=0:A$="in        hops":GOSUB 16000:C=3:A=HO:
     GOSUB 17300
 820 RETURN
 830 REM ===========================
 840 REM Calculate plotting point for NX,NY
 850 IF TX<40 THEN C=TX:R=4+TY:RETURN
 860 C=TX-40:R=15+TY:RETURN
 880 REM ==========================
 890 REM Input values between 0 and 1
 900 GOSUB 12000:IF A<0 OR A>1 THEN GOTO 900
 910 RETURN
 920 REM ==========================
 999 REM :DATA for conc'n, prob'ty of starting & stopping,
     width
1000 DATA .2,.9,.1,5
```

Figure 5.5   Listing for KINWAVE

a chaotic result, but one which still has a well-behaved distribution of bar spacing, etc. The models described below are equally dependent on their stochastic content.

5.5   CHANNEL NETWORK SIMULATIONS

There has been considerable discussion in the geomorphological literature, most significantly by Shreve (1967), about the extent to which randomly generated channel networks are representative of real nets. To test this view, a large number of random nets should be simulated, and the distribution of topological or other properties compared with observed values. In this section, two types of network simulation are presented, together with an algorithm for analysing their Horton–Strahler order structure. The first simulation follows from work initiated by Leopold and Langbein (1962), with streams draining each node of a spatial grid. The second simulation is concerned only with the topology and lengths of each stream link, with no attempt to fit the resulting network on to a basin map; this method is in keeping with Shreve's approach.

   The program for analysing stream orders can also be used to analyse data from a real catchment, measured either in the field or from 1:25,000 maps. Figure 5.7(a) shows a hypothetical network, to illustrate the terminology used and the way to analyse its topology. A *node* is either a point at which the network branches (an *interior* node) or a stream head (*exterior* node). A *link* is the section between adjacent nodes, and is an *exterior* or *interior* link depending on the type of node at its upstream end. The conventional way to describe a network is by walking around it clockwise, starting from the outlet, as indicated by the outer curve in Figure 5.7(a). Each node is described when it is visited in an upstream direction, which is always the first time the node is visited. The nodes are thus visited in the order oabcdefghijk. Each node is described as exterior or interior, and the length

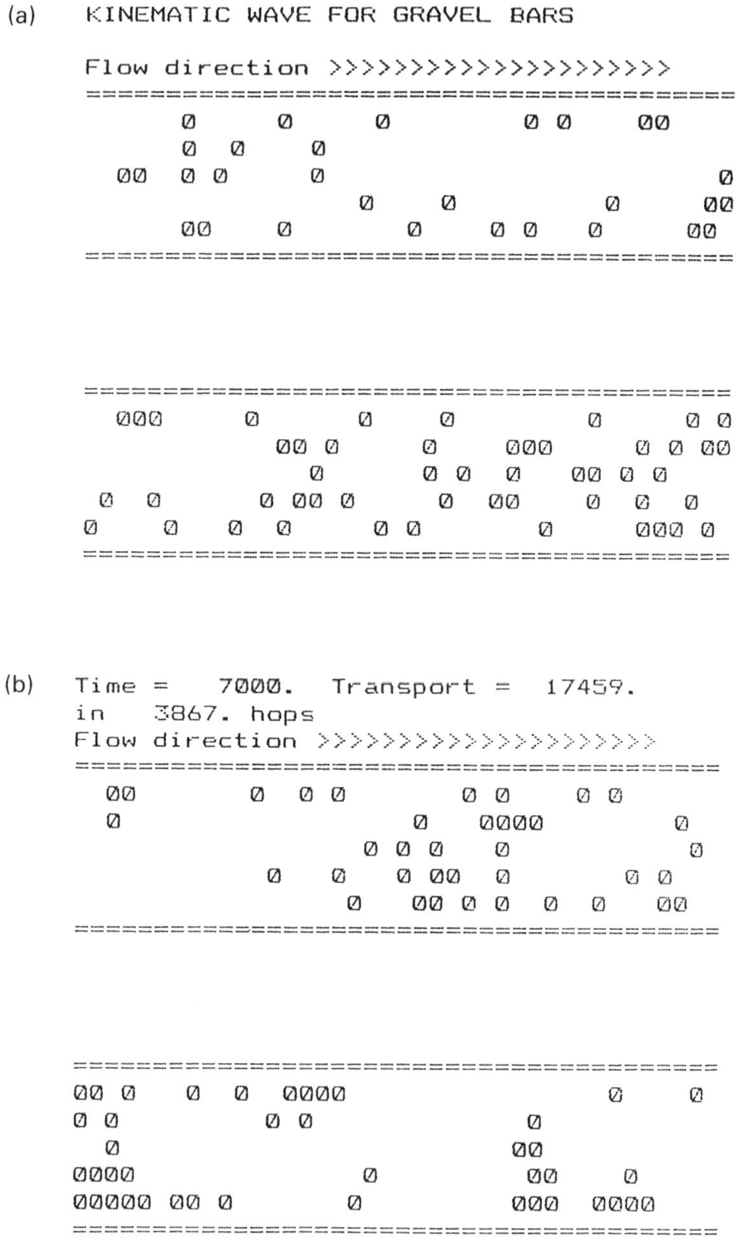
(a)
KINEMATIC WAVE FOR GRAVEL BARS
Flow direction >>>>>>>>>>>>>>>>>>>>>
(b)
Time = 7000. Transport = 17459.
in 3867. hops
Flow direction >>>>>>>>>>>>>>>>>>>>>

Figure 5.6 Sample output for KINWAVE for a channel of width 5 (a) initial random distribution; (b) distribution after 7000 iterations showing some tendency towards bar formation

(a)

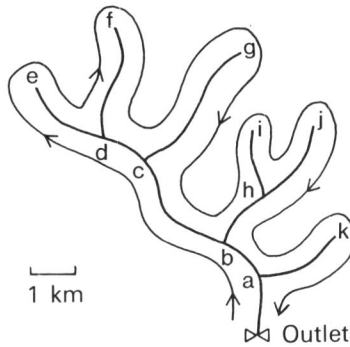

(b) '***' indicates outlet. For a network of magnitude M, the remaining
(2M−1) values correspond to M exterior links and (M−1) interior
links.
In this example Magnitude M=6.

***	+1.2	+0.7	+2·2	+0.9
−1.8	−2.4	−3.2	+1.5	−1.3
−1.7	−1.9			

Figure 5.7 Description of channel network topology and link lengths, obtained by
scanning clockwise, describing each link traversed in an upstream direction; '+'
indicates an interior link; '−' an exterior (fingertip) link of indicated length (a)
diagrammatic scan of links; (b) example sequence of signed lengths

of link downstream from the node can also be uniquely associated with it
(except in the case of the outlet node). The network can now be described
by the sequence of values in Figure 5.7(b). The values are the link lengths,
together with a plus sign for an interior node, or a minus sign for an exterior
node. The initial '+' indicates the outlet node, which is conventionally
regarded as an interior node, but with no link length associated with it. The
next value is +1.2, for an interior link of length 1.2 km flowing downstream
from a, and the remaining ten values refer to each of the nodes b–k in turn.
It may be seen that in this, and in fact for all cases, there is an equal number
of exterior and interior nodes (including the outlet point). This is necessarily
so, since the simplest possible network consists of an outlet and a single
exterior node; and each interior node is a bifurcation which adds one more
exterior node to the network. If the lengths of the links are not of interest,
but only their arrangement, then the sequence can be written simply as a
sequence of 1s for interior nodes and 0s for exterior nodes.
    To analyse a network which has been described by such a sequence of
signed link lengths, the values should be entered as a DATA statement in
the listing for STR AN shown in Figure 5.8, replacing line 3000 which

```
1500 GOSUB 10000:RESTORE: M$="1"
1510 READ MM: DIM S(MM-2),BF(10,10)
1520 READ Q$
1530 FOR I=0 TO MM-2
1540   A$="1": READ A: S(I)=A: IF A<=0 THEN A$="0"
1550   M$=M$+A$
1560   NEXT I
1570 GOSUB 2000: END
2000 REM ================
2010 REM Converts binary string to bifurcation matrix
2020 REM M$=Binary string:BF(10,10)=Bifurcation matrix
2030 REM A$=Order string: Q$= Test string
2040 REM =================
2050 REM Check for valid string & flag errors:
     Transfer 1st order links to A$
2060 L=LEN(M$):A$="":XU=0
2070 SS=0:F=0:FOR I=1 TO L
2080   A=VAL(MID$(M$,I,1)):SS=SS+2*A-1:
       IF SS=0 OR A>1 THEN F=F+1
2090   A$=A$+CHR$(49-7*A):REM '1' or '*'
2100   NEXT I:Q$=M$:IF F<>1 OR SS<>0 THEN
       PRINT "Invalid string":RETURN
2110 REM ================
2120 B$=A$:GOSUB 18000:A$="Order string":GOSUB 24000: A$=B$
2130 REM Search for 1st '100' sequence & update order string
2140 X=1:GOSUB 24000
2150 IF MID$(Q$,X,1)<>"1" THEN X=X+1:GOTO 2150
2160 Y=X+1:REM '1' found
2170 A=VAL(MID$(Q$,Y,1)):IF A=1 THEN X=Y:GOTO 2160
2180 IF A=2 THEN Y=Y+1:GOTO 2170
2190 Z=Y+1:REM 1st '0' found
2200 IF Z>L THEN Q$=Q$:GOTO 2350:REM Network reduced to a
     single link ('10')
2210 A=VAL(MID$(Q$,Z,1)):IF A=1 THEN X=Y:GOTO 2160
2220 IF A=2 THEN Z=Z+1:GOTO 2200
2230 REM Whole'100' sequence found
2240 REM Replace '100' by '022' in Q$
2250 Q$=LEFT$(Q$,X-1)+"0"+MID$(Q$,X+1,Y-X-1)+"2"+
     MID$(Q$,Y+1,Z-Y-1)+"2"+MID$(Q$,Z+1)
2260 U=VAL(MID$(A$,Y,1)):V=VAL(MID$(A$,Z,1)):
     REM Orders of joining links
2270 REM Update bifurcation matrix
2280 BF(U,V)=BF(U,V)+1:BF(V,U)=BF(V,U)+1:BF(U,0)=BF(U,0)+1:
     BF(V,0)=BF(V,0)+1
2290 XU=XU+(XU-U)*(U>XU)+(XU-V)*(V>XU)
2300 IF V>XU THEN XU=V
2310 IF U=V THEN U=U+1:REM Apply ordering rules
2320 IF U<V THEN U=V
2330 A$=LEFT$(A$,X-1)+STR$(U)+MID$(A$,X+1):GOTO 2140
2340 REM Insert new order & look for next '100' sequence
2350 REM =================
2360 REM Print Horton Stream Number Analysis
2400 A$="Order:     1  2  3  4  5  6  7  8  9 10": GOSUB 24000
2450 A$="LINKS    ":FOR  I=1 TO XU
2460   B$="  "+STR$(BF(I,0)): B$=RIGHT$(B$,3): A$=A$+B$:
       NEXT I: GOSUB 24000
2470 A$="STREAMS ":FOR I=1 TO XU
2475   FOR I=1 TO XU:C=INT(24/XU*(I-1))+12
2480     A=0:FOR J=I TO XU:A=A+BF(I,J):NEXT J:IF A=0 THEN A=1
2490     B$="  "+STR$(A): B$=RIGHT$(B$,3):A$=A$+B$: NEXT I:
       GOSUB 24000
```

(*continued*)

```
2500 GOSUB 29000:RETURN
2510 REM ==================
2990 REM Data for link lengths round network, scanned
     clockwise from outlet
2995 REM Begin with NUMBER ; then "+" for lengths: "1" for
     1,0 topologic data
3000 DATA 12,"+",+1.2,+0.7,+2.2,+0.9,-1.8,-2.4,-3.2,+1.5,
     -1.3,-1.7,-1.9
```

Figure 5.8 Listing for network analysis program, STR_AN, which may be used on its own, or merged with either CHNET or CHSTR

begins with the number of nodes (12) and then gives successive values, beginning with '+' for a list of lengths, or '1' for a list of 1s and 0'. When this program is run, the network represented by the data values in line 3000 will be analysed showing a string which is the order associated with each link, and then the number of links and streams of each order. Stream orders are defined according to the normal Horton–Strahler method in which fingertip tributaries are of order one, two first-order streams join to make a second-order stream, etc.

The spatial simulation of drainage nets is readily understood from the example output shown in Figure 5.9(a). Flow drains each square in a 10×10 grid, leaving either to the right or downward. With these two flow directions, which are taken as equally probable, flows may join to form a dendritic network, but there are no opportunities for either closed loops or braiding. A number of distinct networks leave the area along the lower and right-hand edges of the grid. Figure 5.9(b) shows the sequence of signed link lengths for the largest of those networks. The program CHNET listed in Figure 5.10 contains the very simple algorithm to draw the network in lines 10 to 200. The probabilities of the two flow directions may be changed by changing the critical value (0.5) in line 120.

The remainder of the program scans the flow directions starting from each distinct outlet point, to build up a list of values which describes the branching and length of each link, using the method described above. Lines 220 to 330 in the listing identify all the distinct outflow points in turn, and the subroutine in lines 450 to 700 follows each network and assembles its sequence of link lengths. This kind of space-filling simulation can be used to generate distributions of interior and exterior link lengths, which have been suggested as having slightly different distributions. It has also been suggested that competition for drainage area means that successive tributaries to a main stream are likely to be on alternate banks, so that link length distributions are different for *cis*-links (where successive tributaries are on the same bank), as opposed to *trans*-links (successive tributaries on opposite banks). Questions about basin shape can also be asked for this type of simulation, and it may be argued that the requirements of filling a drainage area demand some deviations from strict randomness of topology in Shreve's sense.

110

(a)   Channel net in square of side 20

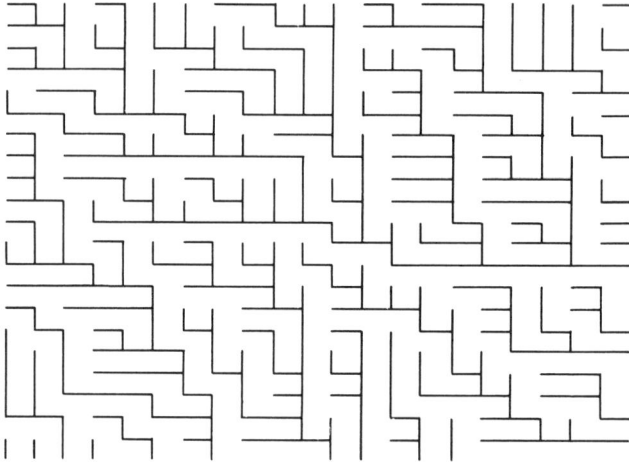

(b)   Network to 20, 12
       ' + ' for Int. link: ' - ' for Ext. link

***	2.	3.	4.	1.	3.	1.
2.	1.	-3.	1.	-4.	-1.	1.
-3.	-1.	-2.	5.	1.	2.	-2.
-7.	-1.	1.	3.	1.	-4.	2.
1.	-1.	-2.	2.	1.	-1.	-1.
1.	-2.	-3.	-2.	-1.	-1.	-2.
1.	2.	-2.	1.	1.	1.	4.
-1.	1.	1.	-3.	-2.	-1.	4.
1.	-4.	-1.	-1.	-1.	-1.	-3.
3.	-2.	1.	-2.	1.	-2.	-1.
1.	-3.	1.	-1.	2.	-2.	1.
1.	-1.	-2.	2.	-3.	1.	1.
-1.	-2.	2.	1.	-1.	-3.	-1.
1.	3.	2.	3.	-1.	-2.	-1.

Figure 5.9 Sample output from CHNET for networks draining a 20×20 grid (a) map of networks; (b) sequence of signed lengths for the largest network

```
10 GOSUB 10000:A=32: DIM BF(10,10)
20 PRINT "Random space-filling Network":C=0
30 C$="Enter size of square to fill: ":R=2:GOSUB 12000
40 A=INT(A):IF A<2 OR A>200 THEN GOTO 30
50 N=A:GOSUB 20000:NX=0:XX=N+1:NY=-N-1:XY=0:GOSUB 21000
60 REM Set up graphics screen
70 DIM N$(N):PRINT "Channel Net in square of side ";N
80 REM =====================
90 REM Main loop
```

(continued)

```
100 FOR IY=0 TO N
110   FOR IX=0 TO N
120     GOSUB 14000:B=0:IF A<.5 THEN B=1
130     N$(IY)=N$(IY)+CHR$(48+B)
140     FG=0:X=IX:Y=-IY:GOSUB 26000
150     IF B=0 THEN X=X+1:GOTO 170
160     Y=Y-1
170     GOSUB 26000:REM Draw link
180     NEXT IX
190   NEXT IY
200 GOSUB 29500:REM Dump Graphics screen
210 REM =========================
220 REM Conversion to Scalar strings
225 GOSUB 900:GOSUB 15000
230 MM=INT(2*N^1.5):DIM S(MM):R=0
240 IY=N:FOR IX=0 TO N-1
260   IF MID$(N$(N),IX+1,1)="1" THEN GOSUB 450
270   NEXT IX
280 IX=N:GOSUB 450
290 FOR IY=N-1 TO 0 STEP -1
310   IF RIGHT$(N$(IY),1)="0" THEN GOSUB 450
320   NEXT IY
330 GOSUB 29000:PRINT:PRINT "Finished":END
340 REM ====================
450 REM Assign to scalar string S()
460 S(0)=1:FOR I=1 TO MM:S(I)=0:NEXT I:M$=""
470 IM=0:X=IX:Y=IY
510 Q=0:IF Y=0 THEN GOTO 530
520 IF MID$(N$(Y-1),X+1,1)="1" THEN Q=2
530 IF X=0 THEN GOTO 550
540 IF MID$(N$(Y),X,1)="0" THEN Q=Q+1
550 ON Q+1 GOTO 560,600,650,700
559 REM Stream head case
560 S(IM)=-S(IM)
570 L=LEN(M$):IF L=0 THEN GOTO 750
580 H$=RIGHT$(M$,2):X=ASC(H$):Y=ASC(RIGHT$(H$,1))
590 M$=LEFT$(M$,L-2):IM=IM+1:S(IM)=1:GOTO 510
599 REM Continuation cases
600 S(IM)-S(IM)+1:X=X-1:GOTO 510
650 S(IM)=S(IM)+1:Y=Y-1:GOTO 510
699 REM Bifurcation case
700 IM=IM+1:M$=M$+CHR$(X)+CHR$(Y-1):X=X-1:S(IM)=1: GOTO 510
710 REM =================
720 REM Printout of scalar string
750 IF R+(IM+2)/8>19 THEN GOSUB 29000:GOSUB 900:GOSUB 18000:
    R=-2
760 C=0:R=R+2:A$="Network to "+STR$(IX)+", "+STR$(IY):
    GOSUB 16000
770 C=0:R=R+1:A$="'+' for Int.link: '-' for Ext. link":
    GOSUB 16000:B=3:C=0:R=R+1
780 A$="***":GOSUB 16000
790 FOR I=0 TO IM
800   C=C+5:IF C>37 THEN C=0:R=R+1
810   A=S(I):GOSUB 17300
820   NEXT I:GOTO 2000
830 REM =================
900 IF F7=2 THEN R=23:C=0:C$="Press RETURN key to continue":
    A$="":GOSUB 11000
910 RETURN
2000 RETURN
```

Figure 5.10   Listing for CHNET

The second network simulation goes directly to the list of signed lengths. It is generated by the program CHSTR listed in Figure 5.11, which has two stochastic elements. The first is a routine for generating a suitable sequence of plus and minus signs. A required stream magnitude (the number of exterior nodes in the network) MG is entered, and the sequence must contain one more '$-$' than '$+$' value. In addition, this condition must not be met at any time before the end of the sequence, of $2MG-1$ values. If this occurs then the network can be shown to have terminated at some magnitude less than MG. It can be shown from the ballot theorem of random walks (see, for example, Feller, 1950, p. 73) that if X is the accumulated excess of '$+$' over '$-$' terms, starting from a value of 1 to represent the outlet, and Y is the number of steps left in the sequence (initially $2MG-1$), the probability of a '$+$' term next is given by:

```
  10 GOSUB 10000:MG=100:LL=10
  20 PRINT "Random Network String":C=0
  30 C$="Enter magnitude of network: ":A=MG:R=1:GOSUB 12000:
     A=INT(A):IF A<1 THEN GOTO 30
  35 MG=A:MM=MG+MG:DIM S(MM)
  40 C$="Enter mean link length: ":A=LL:R=2:GOSUB 12000:
     IF A<=0 THEN GOTO 40
  45 LL=A:PRINT: PRINT "+ for Int.link: - for Ext. link":
     B=3:R=5:C=0 :D=2
  47 A$=" ****":GOSUB 16000
  50 REM =======================
  60 REM Main Routine
  70 X=1:Y=MM-1:IM=0
  80 IF Y=1 THEN P=0:GOTO 100
  90 P=(Y-X)*(X+1)/2/(Y-1)/X
 100 Q=-1:GOSUB 14000:IF A<P THEN Q=1
 110 REM Gamma(2) distribution: see fig 5.1d
 120 GOSUB 14000:RB=A:GOSUB 14100:RA=-A
 130 A=(1+RA)/RB:GOSUB 14100:IF ABS(1-RA/A)>1E-2 THEN RA=A:
     GOTO 130
 140 A=RA*LL/2:REM Random value with mean LL
 150 S(IM)=Q*A:GOSUB 840
 160 IM=IM+1:Y=Y-1:X=X+Q:IF IM<MM-1 THEN GOTO 80
 165 GOSUB 29000
 170 C=12:R=R+1:PRINT:PRINT "Finished "
 180 IM=IM-1
 200 DIM BF(10,10)
 220 GOSUB 2000:END
 830 REM =================
 840 C=C+8:IF C>35 THEN C=0:R=R+1
 845 IF R<23 THEN GOTO 850
 847 PRINT:GOSUB 29000:IF F7=2 THEN GOSUB 900
 848 R=2:GOSUB 15000:PRINT " ";IM;
     " values already printed out of ";MM-1
 850 A=S(IM):GOSUB 17200
 860 RETURN
 880 REM =================
 900 C$="Press RETURN to continue":A$=" ":GOSUB 11000:RETURN
2000 RETURN
```

Figure 5.11 Listing for CHSTR

$$p = (Y-X)(X+1)/[2X(Y-1)] \qquad \textit{if } Y > 1$$
$$p = 0 \qquad \textit{if } Y = 1 \qquad (5.9)$$

A suitable random value of Q is generated in lines 70 to 100.

The second stochastic element in the model is for the distribution of link lengths. These have been generated using a gamma(2) function using, in lines 110 to 140, the routine taken from Figure 5.1(d). An example of the output from this program is shown in Figure 5.12(a) for a net of magnitude 30. This gamma distribution has been found to give the best empirical fit to many link length data, although alternative distributions may readily be substituted. Different distributions or means could also be used for interior and exterior links. These modifications are left as an exercise for the reader.

The main routine ends at line 170. Lines 180 to 200 convert the topology of the network to the string M$ of 1s to represent interior nodes ('+'s) and 0s to represent exterior nodes ('−'s).

The networks produced by either of the random simulation programs, CHNET or CHSTR, may be merged with the network analysis program STR_AN (using *EXEC STRANEX on the BBC micro). The analysis may

---

(a) Random Network String
Enter magnitude of network: 30
Enter mean link length: 10

+ for Int. link: − for Ext. link

****	15.08	−14.20	3.73	11.25
11.47	13.09	−9.55	−5.58	−0.16
17.08	30.34	28.35	−8.06	3.54
5.59	4.85	−1.04	−22.48	−1.86
−12.98	12.26	10.28	−11.88	17.76
9.28	−7.99	−5.25	−10.88	−12.76
10.88	−4.55	17.37	6.62	2.56
7.84	9.42	−5.79	−10.84	−5.57
16.33	−0.92	−6.59	−9.68	17.01
8.98	34.66	−10.47	2.26	−0.91
−6.32	4.08	3.69	−3.40	−5.78
2.30	−17.70	−7.57	−15.99	−14.58

(b) Binary and Order strings
1101111000111011100001101100001011111000
10001110100110010000

*41444221114321222111122122111141433 22111
21113321211321121111

Figure 5.12 Sample output from CHSTR and STR_AN (a) sequence of signed lengths for a network of magnitude 30; (b) analysis of topology to give Horton–Strahler order for each link, and total numbers of links and streams by order

be used in its present form for magnitudes of up to 127 on the BBC micro. Figure 5.12(b) shows an example output from this analysis for the network shown in Figure 5.12(a); its topology is shown in Figure 6.9(a). An analysis of stream lengths can readily be made by associating the order of each link with its length shown in Figure 5.12(a).

These channel network simulations again rely totally on a stochastic input, and the 'average' case of a deterministic model would consist of a simple regular pattern. The expected distributions of topological length and area relationships cannot be derived through any simple analysis, if at all, so that these simulations are vital to the identification of unusual real networks which show, for example, significant structural control.

5.6 A FRACTAL SURFACE

The last detailed example of a stochastic model is for a surface generated within a two-dimensional square frame. As with the one-dimensional sequences generated in Section 5.2, the problem is to generate a sequence of values which are partly dependent on neighbouring values and partly independent. The fractal surface used here has the property of having dependence on its neighbours at all scales, as opposed to the Markovian and running mean methods which have a definite range of dependence. It has been described as having a fractional dimension of somewhere between 2 (for a normal smooth surface) and 3 for a 'surface' which completely fills a volume. At intermediate dimensions, the surface area, if examined in sufficient detail, is infinite, although any practical realization of the fractal is limited by available computer memory.

Mandelbrot's (1977) book provides a mathematical introduction to fractals, and is illustrated with many visually stunning examples. The present simulation is much more limited, but still of use in practical models. Many actual surfaces and sequences in environmental science have been shown to have fractal-like properties, so that these models may be used to compare with real distributions. One of the earliest examples of a fractal curve was the coastline. Richardson (1961) showed that if its length was measured with a progressively shorter measuring stick, then the total length increased, apparently indefinitely, as an inverse power of the unit of measurement. This is in contrast to measurement of a smooth curve like a circle, for which the measured length increases towards a definite limit equal to the true circumference. The exponent of the rate of increase may be thought of as an 'excess dimension' over the normal value of 1.0 for a curve. Coastlines and river courses are among the examples of fractal curves in the landscape, usually with an excess dimension of about 0.2 to 0.3. It may be argued that low excess dimensions are found where diffusive transport processes remove irregularities, and high excess dimensions where these processes are absent. Thus on a hillside, excess dimension is greater along the contour than downslope.

As well as a basis for comparison, a fractal surface or sequence may be used as input to a deterministic model, in much the same way as for the random sequences discussed in Section 5.2. The surface produced by the program listed here (Figure 5.13) might be treated as the initial surface, or at least the irregular component of the initial surface on which hillslope and soil processes may be simulated deterministically. Small initial irregularities provide the flow convergence and divergence which is required to lead to the eventual formation of major hills and valleys.

Construction of the fractal surface is an approximate procedure, relying on successively halving the interval between points, and adding a random component to the value interpolated between the four nearest points. The variance of the random component is progressively reduced with the distance to the neighbouring points, as a power function which depends on the excess dimension. When the distance over which interpolation is made is x, then the standard deviation of the random component is:

$$SD \propto x^{(ED-1)} (1 - 2^{-2ED})^{1/2} \tag{5.10}$$

where ED is the excess dimension required. The first part of this expression is the power function of distance, and the second is a correction to eliminate the variance components correctly from the four neighbouring points. It may be seen that for an excess dimension of zero, there is no random component, so that the process is one of simple interpolation. For an excess dimension of 0.5, the standard deviation is proportional to $x^{-1/2}$, corresponding to a normal distribution of residuals. For an excess dimension of 1.0, the standard deviation remains constant, corresponding to a simple random walk. It may be seen that as interpolation continues, this 'surface' for ED $= 1$ fills more and more of the volume, so that it is not unreasonable to associate it with a total dimension of ED$+2 = 3$.

The program begins by entering a central elevation from the keyboard, and setting the marginal elevations all to zero. Figure 5.14 illustrates successive stages of interpolation on a 32×32 grid. In Figure 5.14(a) the central point is shown by '0'. On the first pass, the four points marked '1' are interpolated, from the existing values at the corners of the square indicated. Thus (8,8) is interpolated from (0,0), (0,16), (16,0) and (16,16), etc. The interpolation distances are 8 in the x and y directions, but $128^{1/2}$ measured in a straight line. At the second pass, the points '2' are interpolated as shown from the existing corners of the surrounding diamond. Thus (16,24) is interpolated from (16,16), (16,32), (8,24) and (24,24) as shown in Figure 5.14(a). The interpolation distance is now 8. In Figure 5.14(b) the already interpolated points are again shown by '0' and the two passes at this scale are again shown as '1' and '2' respectively. Thus at each pass, the interpolation distance is reduced in the ratio of $2^{1/2}$. This procedure can continue as long as desired, or as long as computer memory permits.

Figure 5.15 shows the output from FRACTAL. In Figure 5.15(a) the edges of the black and white areas define contours at a vertical interval of

```
 10 GOSUB 10000:PP=5:ED=.2:SF=20:CH=50:A=PP:
    PRINT "        FRACTAL SURFACE"
 20 C=0:R=2:C$="Enter number of stages: ":GOSUB 12000:
    A=INT(A):IF A<2 OR A>6 THEN GOTO 20
 30 PP=A:A=ED
 40 R=3:C$="Excess dimension (0-1): ":GOSUB 12000:
    IF A<0 OR A>1 THEN GOTO 40
 50 ED=A:A=SF
 60 R=4:C$="Scale factor: ":GOSUB 12000:
    IF SF<0 OR SF>100 THEN GOTO 60
 70 SF=A:A=CH
 80 R=5:C$="Centre Height: ":GOSUB 12000:
    IF A>100 OR A<-20 THEN GOTO 80
 90 CH=A:SS=2^PP:DIM H$(SS)
100 A$=CHR$(50):FOR I=1 TO PP:A$=A$+A$:NEXT I
105 A$=A$+CHR$(50)
110 FOR I=0 TO SS:H$(I)=A$:NEXT I
120 D=1-ED:HH=2^(-D/2):KA=SQR(1-2^(2*D-2)):KS=KA*SF:
    L=SS/4:QF=SQR(12)
130 GOSUB 20000:NX=0:XX=3*SS/2:NY=0:XY=SS+1:GOSUB 21000:
    REM Open Graphics screen
135 FG=0:X=0:Y=0:GOSUB 26000:X=SS+1:GOSUB 26000:Y=SS+1:
    GOSUB 26000:X=0:GOSUB 26000:Y=0:GOSUB 26000:
    REM Draw frame
140 X=SS/2:Y=X:Z=CH:GOSUB '1000:C=0:R=1:REM Centre point
145 GOSUB 18000:PRINT "        FRACTAL SURFACE"
150 REM ==================
160 REM Main loop
170 FOR U=1 TO PP-1
180   FOR F=0 TO 1
185     C=0:R=1:A$="Every "+STR$(L)+": Pass "+
        STR$(F+1)+"   ":GOSUB 16000
190     X=L*(1+F):Y=L
200     REM Inner loop
210     GOSUB 700:REM Calculate Height
220     GOSUB 1000: REM Set & draw point
230     X=X+L+L:IF X>=SS THEN Y=Y+L*(2-F):X=X-SS+L*F
240     IF Y<SS THEN GOTO 210
250     REM End inner loop
260     REM =============
270     KS=KS*HH:NEXT F
280   L=L/2:NEXT U
290 REM ================
300 GOSUB 29500:GOSUB 600
310 REM Draw cross-sections
320 GOSUB 28000:NX=0:XX=SS:NY=-16:XY=96:GOSUB 21000:
    SY=16:GOSUB 22000
330 C=0:R=0:A$="Cross sections":GOSUB 16000:PRINT
340 GOTO 390
350 A$=H$(IY):FG=0
360 FOR X=0 TO SS
370   Y=ASC(MID$(A$,X+1))-50:GOSUB 26000
380   NEXT X
390 PRINT "Enter Y-value: ";:INPUT A$
400 IF A$="" THEN GOTO 300
410 IY=VAL(A$):IF IY<1 OR IY>SS-1 THEN GOTO 390
420 GOTO 350
600 REM ================
610 IF F7<>2 THEN RETURN
```

(continued)

```
 620 C$="Press RETURN to continue ":A$="":C=0:R=0:
     GOSUB 18000:GOSUB 11000
 630 RETURN
 680 REM =================
 690 REM Interpolate mean height
 700 IF F=1 THEN GOTO 720
 710 Z=ASC(MID$(H$(Y-L),X+1-L))+ASC(MID$(H$(Y-L),X+1+L))+
     ASC(MID$(H$(Y+L),X+1-L))+ASC(MID$(H$(Y+L),X+1+L)):
     GOTO 730
 720 Z=ASC(MID$(H$(Y),X+1-L))+ASC(MID$(H$(Y),X+1+L))+
     ASC(MID$(H$(Y-L),X+1))+ASC(MID$(H$(Y+L),X+1))
 730 Z=Z/4-50:REM Interpolated value
 740 GOSUB 14000:R=(A-.5)*QF:REM Random
 750 Z=Z+KS*R:RETURN
 990 REM =====================
 995 REM Set point on screen
1000 CR=Z+50:CR=CR+CR*(CR<0)+(CR-255)*(CR>255)
1010 REM Set value in array
1020 H$(Y)=LEFT$(H$(Y),X)+CHR$(CR)+MID$(H$(Y),X+2)
1040 REM Plot point on screen
1050 IF Z/32 -INT(Z/32)<.5 THEN RETURN
1060 XA=X+1:YA=Y+1:GOSUB 27000:RETURN
1070 REM ==================
```

Figure 5.13   Listing for FRACTAL

16. The central area represents a peak and the marginal black areas are depressed below the zero margin. Figure 5.15(b) shows four sections across the feature. The reader may wish to experiment with different values for the parameters, ED, CH (the height of the central peak) and SF (a scale factor which determines the initial importance of the random component. The grid size is determined by the number of stages, PP, which selects a grid of size 2^{PP}. For the BBC 'B' micro, a maximum value of PP = 6 (i.e. a 64×64 grid) is achievable. Figure 5.15 illustrates a run for PP = 5 (i.e. 32×32).

The program may be simplified to produce a one-dimensional sequence, which could be of up to 2^{12} = 4096 elements. Such fractal sequences are thought to be appropriate to simulate long-term hydrological and meteorological sequences where the amounts of climatic or other external change are not well understood.

5.7   OTHER EXAMPLES OF STOCHASTIC MODELS

There are many other examples of possible uses for stochastic models. This section briefly outlines two further examples without providing listings. In some cases, it is not difficult to develop an outline program, and the reader may wish to attempt one or more. The topic discussed here are flood frequency distributions and diffusion.

The forecasting of flood frequency distributions for ungauged catchments may be approached either on a black box basis or through simulation. The

118

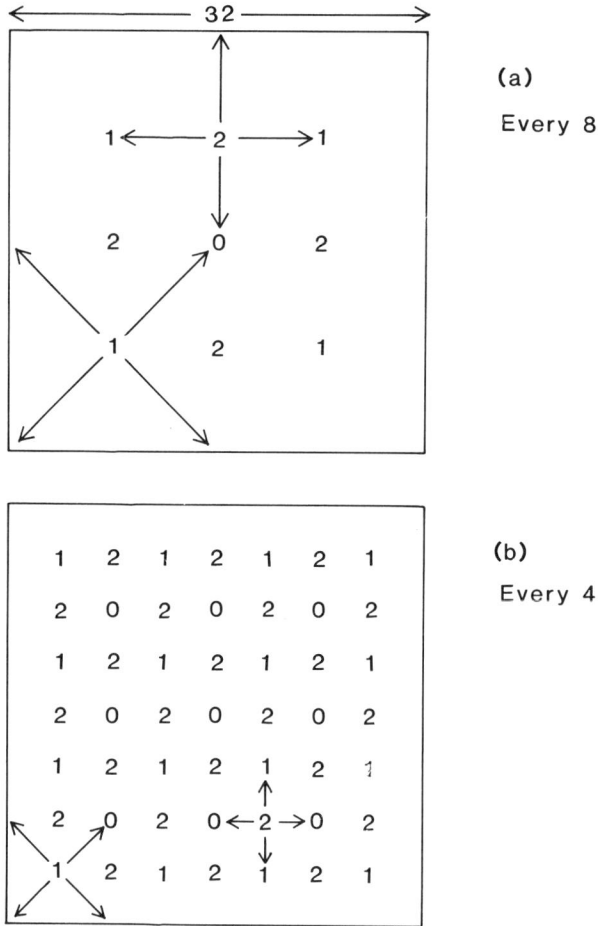

Figure 5.14 The sequence of successive interpolation used to produce a 32×32 fractal surface (a) interpolating at intervals of 8; (b) interpolating at intervals of 4

black box approach regresses basin characteristics among gauged catchments, so that the parameters of the frequency distribution may be estimated for the ungauged catchment. The simulation approach works through some form of catchment model which is valid for the ungauged catchment, together with stochastic rainfall inputs drawn from known rainfall distributions. Storms provide the high flows of interest while most models work on a time increment of a few minutes to a few hours. It is therefore efficient not to generate a sequence of hourly rainfalls over a century or more, but to generate separate random values for storm rainfall and intensity, for antecedent moisture content and for the interval between storms. The model

FRACTAL SURFACE
Every 1: Pass 2

(a)

Cross sections
Enter Y-value: ? 4
Enter Y-value: ? 8
Enter Y-value: ? 12
Enter Y-value: ? 16
Enter Y-value: ?

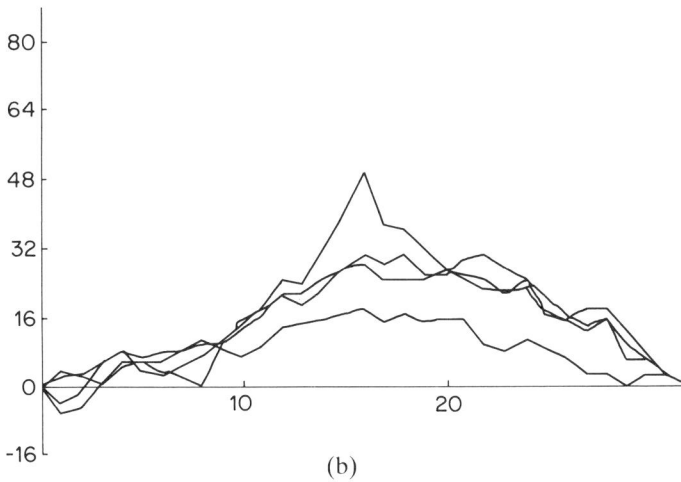

(b)

Figure 5.15   Sample output from FRACTAL for a 32×32 surface (a) contours at a vertical interval of 16; (b) sections from left to right across the surface, at $y=4$, 8, 12 and 16

then needs to be run only for each storm period, keeping run times to a manageable level. Work along these lines (e.g. Beven, 1986) is able to give frequency curves which agree well with those for gauged catchments.

The equation for soil creep down a hillside is mathematically identical to the diffusion equation. Diffusion may equally be represented as a stochastic process, in which particles move freely in all directions, subject to the constraint of available space to move into. Indeed it has been argued that this is one significant mechanism of soil creep on the soil particle scale. Thus the process of hillslope development by soil creep or similar process may be modelled as a stochastic process, instead of in the usual way as a deterministic process as set out in Chapter 4. Where the number of particles moving is very large, as perhaps for soil creep, the stochastic approach is rather inefficient and its outcome shows little spread about the average. The routine involved shows strong similarities with the kinematic wave approach discussed earlier, although material is usually gradually degrading and levelling out, so that the recirculating element is absent.

Although this chapter has only scratched the surface of stochastic modelling, the outline programs and models may give some guidelines about the potential for this type of modelling. As computer power continues to expand and become more widespread, the disadvantages of slightly greater complexity and the need for many replications seems likely to be more than offset by the greater testability of the outcomes. Because computer simulations are much cheaper than field experiments, and usually much quicker too in the field sciences, it makes sense to increase the number of computer replications to minimize field programmes and to squeeze a high level of interpretation from them. The development of stochastic models has an important role to play in this process.

5.8 REFERENCES

Beven, K. J. (1986) Runoff production and flood frequency in catchments of order *n*: an alternative approach. In V. K. Gupta, I. Rodriguez-Iturbe and E. F. Wood (eds) *Scale Problems in Hydrology: Runoff Generation and Basin Response* Reidel, Dordrecht, 107–32.

Feller, W. (1950) *An Introduction to Probability Theory and its Applications*, Wiley, New York.

Langbein, W. B. and Leopold, L. B. (1968) River channel bars and dunes – theory of kinematic waves, *U.S. Geol. Survey, Professional Paper*, 422L.

Leopold, L. B. and Langbein, W. B. (1962) The concept of entropy in landscape evolution, *U.S. Geol. Survey, Professional Paper*, 500A.

MacArthur, R. H. and Wilson, E. O. (1967) *The Theory of Island Biogeography*, Princeton University Press, Princeton, N.J.

Mandelbrot, B. B. (1977) *Fractals: Form Chance and Dimension*, Freeman, San Francisco.

Richardson, L. F. (1961) The problem of contiguity: an appendix of statistics of deadly quarrels, *General Systems Yearbook*, **6**, 139–87.

Shreve, R. L. (1967) Infinite topologically random channel networks, *J. Geology*, **75**, 178–86.

Part II

CHAPTER 6

Model Formulation and Construction

6.1 OVERVIEW

The process of planning and putting together a working model is described in this chapter. At this stage we must take for granted many of the finer points which are discussed in Chapters 7 and 8. The construction of a model begins by defining as our aim the environmental variables which are to be forecast or simulated. This overall aim must then be refined by defining the relevant parts of the controlling environmental system. A series of important decisions must be made about the level of detail which is worth including and about the general style of approach you wish to follow. These issues are discussed in Chapter 8. Here we will assume that provisional decisions have been taken. If they are not justified by the performance of the completed model, then a new start may be necessary.

Another important set of decisions must also be made about how the final model is to be calibrated, verified and used (Chapter 7). Although less involved with the content of the model program, attention must be paid to these needs, so that the forecasting 'core' of the model produces output which is compatible with them. In many cases, routines for calibration and verification will ultimately need to be incorporated into the final computer program. In other cases it may be sufficient to produce disc or other files of forecast data values in a form which can be 'read' and digested by special purpose or standard packages/programs for statistical and other appropriate analyses. In the final application of the forecasting program, it may or may not be necessary to include the routines to calibrate the model parameters. In simple cases, the forecasting core may be all that is needed. In final application, however, the form of output which is most helpful to the user is likely to be somewhat different from that which is needed at the development stage. It is assumed for now that decisions have been reached on all these points of general strategy. The remaining problem of model construction is to define a routine or algorithm to represent the processes or transfer functions in a logical and sequential way, and to convert them into a correctly working BASIC program. This is our concern here.

Figure 1.1 summarizes the main steps in creating a computer model. We will go through them in turn with examples, mainly drawn from previous chapters, to illustrate the stages and to bring out some points of detail. The first step is to define a procedure which leads to the desired forecast, in logical or mathematical terms. In many cases it is helpful to proceed through a systems diagram, or a system of equations, perhaps together with some kind of physical diagram of the forces, flows, etc. present. The second stage is to put the operations involved into a definite sequence, since the computer must work in this way, even when representing actions which are, in reality, simultaneous. The result of this stage may be represented by a flow diagram. Next the type of input and output should be considered, and its operations included in the flow diagram. It is then possible to convert the flow diagram into a computer program in BASIC (or any other appropriate language). Finally, the program needs to be tested for ability to run at all, and then ability to produce the desired forecast values. When all these stages are complete, we may wish to refine the program further in the light of its performance.

6.2 DEFINING AN ALGORITHM

The first step in constructing a model, and one of the most difficult in any but elementary cases, is to define a procedure or *algorithm* which gives the required forecast, expressed in a mathematical or logical form. There are at least three kinds of layout, some or all of which may be useful in a particular case: a systems diagram, a physical diagram and a set of equations.

A systems diagram is usually most helpful where there are a number of similar or dissimilar components with interactions between them. It is also nearly always helpful in determining an appropriate system of interest for a model, and for systems constrained by mass or energy balances. A systems diagram, composed of 'boxes and arrows', normally distinguishes between *system variables, external variables* and *linkages*. System variables represent the states of system 'objects' which are both influenced by, and themselves influence other system objects (and the variables associated with them). External input variables influence objects in the systems and their variables, but are not themselves influenced by them. In this context, input and output refer to the direction of influence rather than to the sign of the flows to or from them. Thus an external input may refer, for example to *removal* of evaporation at a fixed rate. External output variables are influenced by system objects, but do not influence them. Linkages represent the influences between system objects, or between system objects and external variables, through which the values of system variables are changed.

The choice of a suitable system is entirely in the hands of the modeller, allowing him or her to choose the minimum set of objects to achieve an adequate forecast. The decisions involved are discussed in Chapter 7. A good example of fairly drastic system minimization is illustrated in Figure

3.1, in which a conceptual hydrological system is reduced, for the purposes of the model, to a single system variable (soil water), with two external inputs (infiltration and evapo-transpiration) and two external outputs (saturation overland flow and throughflow), which are the targets of the forecast. The resulting model may be conceived as an independent entity, but might also be one component within a model of the larger hydrological system shown in the figure.

Particularly where the system is constrained by a mass or energy balance, the next important stage is to identify the 'currency' of the system. For the hydrological example of Figure 3.1, the currency is plainly quantities of water, held in physically identifiable stores which are the variables associated with each 'box' or object in the diagram; while the linkage 'arrows' represent transfers or flows (or flow rates in a time based model) between the objects.

Figure 6.1 shows a flow diagram for the program 'GROWVEG' described in Section 4.3.2. There are two system objects, representing live vegetation and organic soil. The currency of the model is clearly some form of biomass, but the model might be successfully run with several alternative versions of the biomass. We might use total dry weight of organic material (i.e. carbohydrate), total organic carbon, or total organic potassium, etc. Since the plant community must combine its nutrients in certain proportions to its best advantage, a case could be made for running parallel models for two or more of these currencies, for example for carbon and a critical inorganic nutrient element, say potassium. The system might therefore have more than one currency. Where we have a number of systems running in parallel, they must connect at some point. Our final result may simply be the sum (perhaps with a suitable conversion factor) of the outputs from each system. In the vegetation example, the actual rate of biomass growth might be determined by the lower of two rates forecast from the carbon and potassium needs. More generally, there may be a number of cross connections between the systems.

The next step in developing an algorithm from a systems diagram is to establish mathematical or logical expressions to represent the values (or rates) of the flows (arrows) in terms of constant parameters and some or all of the system variables, using the example of Figure 6.1, and using total dry weight of carbohydrate as the currency. Figure 6.1(b) shows this stage with the hypothetical functions used in the model. Clearly, in a working model, it is important to make every effort to make these functions as precise as possible within the constraints of the model framework. F(month), G(month) represent rate factors which vary systematically with month of the year, through the seasonal pattern of climate and through biological clocks. These functions need to be put into a more definite form to make the model operational. The model can then be represented by a set of equations, representing the mass balance constraints as storage equations for each system object; and the process equations which are the functional equations for flow rates in each link.

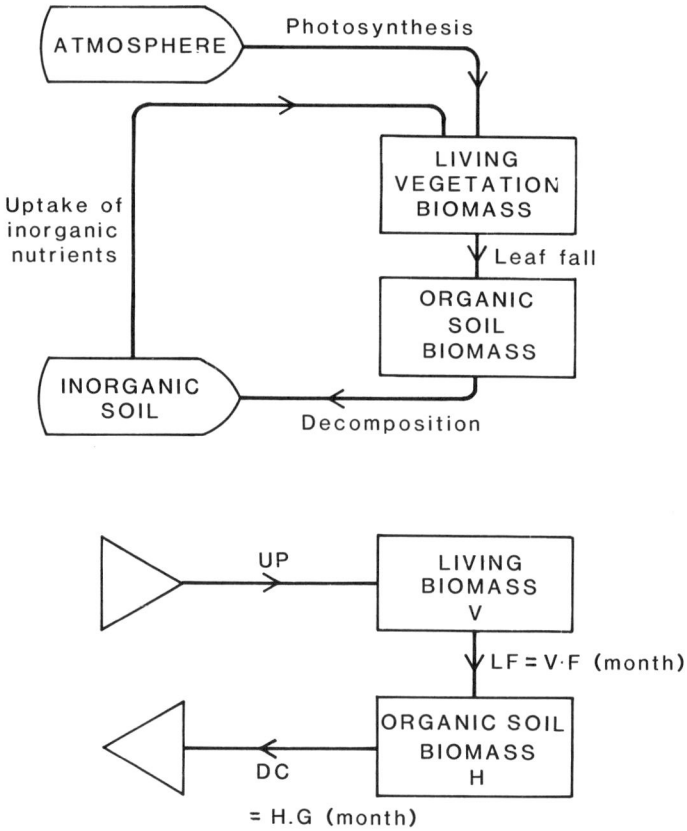

Figure 6.1   Flow diagrams for GROWVEG

The set of equations for this example are:

Mass balance

$$V' = V + (UP - LF)\Delta M$$

$$H' = H + (LF - DC)\Delta M \qquad (6.1)$$

$$M' = M + \Delta M \text{ (returning to month 0 from 12)}$$

Process

$$LF = V\,F(M)$$

$$DC = H\,G(M)$$

where M is the month of the year (0 to 12), and M', V', H' represent updated values.

Figures 2.5(a), 3.8 and 3.11 are examples of physical diagrams which assist in model formulation. Figure 2.5(a) is a schematic map showing the contribution from four sections of a catchment, in time area analysis. It helps to lead to the summation equation for basin discharge. Figure 3.8 shows a balance of forces, which may be resolved in two directions at right angles to one another to give process equations which are relevant to the forecast of grain movement probabilities. Figure 3.11 shows the geometry of a simple wedge landslide, and helps in analysing the masses and forces involved in calculating whether the wedge is stable. A wide range of physical, or in some cases chemical, diagrams may be useful, especially in defining process relationships.

In the very simplest kinds of models, especially black box models, the forecast is simply expressed as one or a series of equations. The first part of the program UNIT in Figure 2.8, as far as line 150, is of this form. The peak discharge (QP), time to peak (TP) and time to revert to base flow (TB) are calculated using three equations (at lines 140 and 150) from a series of input values. In this section of the program there is no need for a systems or other diagram. The set of equations define the forecast values needed.

Many other problems are defined directly through one or a series of equations. Where the model forecasts change in a system over either space or time, then the equations may well be differential or difference equations. Without going into the wide variety of mathematical and computational methods for solution of equations, it is worth pointing out three common problems which occur in writing simple models. These are implicit equations, sets of simultaneous equations and differential equations.

6.2.1 Implicit equations

Implicit equations are those without an explicit solution for the variable required. For example the simple-looking equation

$$y = x - \ln x \qquad (6.2)$$

can readily give y from a value of x, but not x from y. There are three simple ways of solving for x, each of which is successful in some cases. In the first two methods, the first step is to change the expression into one which should be zero, in this case:

$$z = x - \ln x - y \qquad (6.3)$$

The first method is a simple search procedure, like that included in the landslide stability analysis, 'SLIDE', listed in Figure 3.12. An appropriate subroutine is listed as the subroutine from lines 699 to 750 of the program fragment ITER listed in Figure 6.2. The search is made at regular intervals over a pre-defined range, LL to UL, evaluating the expression for z at equally spaced points. If one of the points gives an exact zero, the program exits with this solution. Otherwise the search continues until there is a change in sign, from plus to minus, or vice versa. When the change occurs, the search is repeated in greater detail over the critical interval, and so on until

128

```
  5 MODE 0:PRINT "Comparison of methods for finding roots
    for X": PRINT "of the Equation Y=X-LN(X) for different
    values of Y":PRINT:PRINT"Edit Lines 2000 to 2010 for
    another function"
 10 ER=1E-5 :PRINT:PRINT
 15 PRINT "SYSTEMATIC SEARCH begins"
 20 FOR Y=2 TO 21
 30   LL=1:UL=30:GOSUB 700:NEXT Y
 40 PRINT "SYSTEMATIC SEARCH done"
 50 PRINT: PRINT "NEWTON begins"
 60 FOR Y=2 TO 21
 70   X=Y:GOSUB 850:NEXT Y
 80 PRINT "NEWTON done":PRINT
 90 PRINT "ITERATION begins"
100 FOR Y= 2 TO 21
110   X=Y:GOSUB 1000:NEXT Y
120 PRINT "ITERATION done"
130 END
490 REM =====================
500 PRINT "For Y = ";Y;", Estimated root is ";X:RETURN
690 REM ========================
699 REM Systematic search method
700 F=0:DX=(UL-LL)/2:FOR X=LL TO UL STEP DX
710   NZ=FNZ(X):IF X=LL THEN GOTO 730
720   IF NZ*OZ<=0 THEN F=1:NX=X:X=UL:IF NZ=0 THEN DX=0
730   OZ=NZ:NEXT X:IF F=0 THEN PRINT "None found" :RETURN
740 IF ABS(DX) >ER THEN LL=NX-DX:UL=NX:GOTO 700
750 X=NX:GOTO 500
840 REM ========================
849 REM Newton's method
850 Z=FNZ(X):DZ=(FNZ(X+1E-4)-Z)/1E-4
860 DX=Z/DZ: IF ABS(DX)>ER THEN X=X-DX:GOTO 850
870 GOTO 500
990 REM ========================
999 REM Iterative method
1000 NX=FNU(X):IF ABS(NX-X)>ER THEN X=NX:GOTO 1000
1100 GOTO 500
2000 DEF FNU(X)=LN(X)+Y
2100 DEF FNZ(X)=X-LN(X)-Y
```

Figure 6.2   Listing for ITER, illustrating three methods for solving implicit equations: systematic search at lines 699–750; Newton's method at lines 840 to 870; and iterative method at line 1000

a zero is found within an x-interval of less than a specified error, ER. This method can readily be adapted to give all the zeros within the given range.

The second method, due to Newton, starts with a first guess of a value for x. It then draws the tangent to the x, z curve down to meet the z-axis. This is nearly always an improved estimate of the required value of x. Again the process is repeated until it gives the required degree of accuracy. Lines 840 to 870 give a skeleton listing for this method.

The third method (line 1000) is less general in its application, and requires examination of the particular function for which a value is required. In this case the original function needs to be rearranged with the required variable x on the left of the equation. For our example this could be done in two ways:

$$
\left.
\begin{aligned}
x &= \ln x + y \\
x &= \exp(x-y)
\end{aligned}
\right\} \tag{6.4}
$$

In each case, a value of x entered on the right-hand side of the equation gives another estimate of x by substitution. If the equation has been rearranged in a useful way, then the process is one of convergence on a solution. This occurs for the first form, while use of the second tends to diverge. This choice might be inferred from the fact that $\ln x$ tends to have less range than x, while $\exp(x)$ has more range. An example of this method may be seen in Figure 5.1(d) to derive a random value drawn from a gamma(2) distribution.

As may be seen by experiment, there is not much to choose between the second and third methods for this example, while the search method is strikingly slower. It does, however, have advantages for functions with rapid or abrupt changes in gradient. Only the search method can provide more than a single value for x. A more efficient procedure is to locate roots roughly through a search and to home in accurately on each using one of the faster methods.

6.2.2 Sets of simultaneous equations

In a number of systems, the set of equations which describe the system are not sequential, in the sense that they may be solved explicitly, one at a time, but are interdependent. Where the set of equations is linear, it is most effectively solved by matrix methods. Thus three equations may be written as:

$$
\begin{aligned}
a_{11}x_1 + a_{12}x_2 + a_{13}x_3 &= y_1 \\
a_{21}x_1 + a_{22}x_2 + a_{23}x_3 &= y_2 \\
a_{31}x_1 + a_{32}x_2 + a_{33}x_3 &= y_3
\end{aligned} \tag{6.5}
$$

Alternatively they may be written in matrix notation as:

$$
Ax = y \tag{6.6}
$$

where A is now the matrix with element at column i, row j of a_{ij}, and x, y refer to the 1×3 arrays of values (column vectors). It is clear that, in principle, the inverse matrix A^{-1} is required to solve the equations, giving

$$
x = A^{-1}Ax = A^{-1}y \tag{6.7}
$$

Without going into the mathematics involved, the listing in Figure 6.3 for MATINV performs this operation, inverting an $N \times N$ matrix A using the subroutine starting at line 8000. The matrix is represented by an $N \times N$ array, $A(I,J)$, and inverts it into itself so that the original array is lost. The first part of the program may be run to show how this inversion subroutine may be used to solve a set of equations like those in equation (6.5) above. As listed it solves four equations, but if the data value at line 200 is changed from 4, the program data will solve any number from one to six simultaneous equations.

Where the set of equations is not linear, the matrix inversion method breaks down, and it is usually best to look for a less formal method for each particular case. An iterative procedure is often successful with two variables, though it becomes less effective with more variables. For two variables and two equations, the iteration begins with an estimate of one variable, say x. Then one of the equations is used to estimate y from this value of x. The other equation is then used to estimate a new value of x from the y-value. This process is repeated until it converges on a point, or is clearly either divergent or cyclic. Some of the possibilities may be seen from the sketches in Figure 6.4. The path of successive approximations may spiral around the solution, either towards it or away from it, or move in or out in a stepped pattern. At each stage the estimation of y from x or vice versa may be a straightforward explicit function, or may be an implicit equation, requiring the methods outlined above. You will need to experiment with the choice of method and which equation to use in which direction.

6.2.3 Differential equations

There is a large body of literature on the computational solution of differential equations. Those pursuing the matter in any depth would be advised to refer to specialist texts. The suggestions here will, however, allow an effective start in solving ordinary and partial differential equations. Simple examples have been quoted in the text as equations (3.2), (3.18) and (3.19). Equations (6.1) are an example of a set which have been written in difference form, but are in fact approximate statements of the differential equations:

$$\frac{dH}{dM} = UP - H\,F(M)$$

$$\frac{dV}{dM} = H\,F(M) - V\,G(M)$$

$$(6.8)$$

with notation as before. In fact the difference equations (6.1) are both the basis for common sense projection at current rates and the formal first approximation to the differential equations (6.7). The differential form, while exact, is not computable without approximation. The formal statement is:

$$\frac{dH}{dM} \simeq \frac{\text{change in } H}{\text{change in } M} = \frac{\Delta H}{\Delta M} = \frac{H' - H}{\Delta M} \qquad (6.9)$$

```
   10 CLS: RESTORE: READ N:DIM A(N,N+1)
   20 PRINT "Example solving 4 linear equations":PRINT
   25 PRINT "Modify DATA at lines 190 to 250 to change
      equations":PRINT
   30 FOR I=1 TO N:FOR J=1 TO N+1:READ A(I,J):NEXT J: NEXT I
   40 FOR I=1 TO N
   50   A$="":FOR J=1 TO N
   60     A$=A$+STR$(A(I,J))+CHR$(122-N+J):
          IF J<N THEN A$=A$+"+"
   70     NEXT J:A$=A$+"="+STR$(A(I,N+1)):PRINT A$: NEXT I
   80 PRINT: GOSUB 8000: REM Invert Matrix
   85 PRINT "Solution is: ": PRINT
   90 FOR I=1 TO N
  100   A=0: FOR J=1 TO N
  110     A=A+A(I,J)*A(J,N+1):NEXT J
  120     PRINT CHR$(122-N+I)+" = "+STR$(A): NEXT I
  130 END
  190 REM Data on Number of unknowns
  200 DATA 4
  210 REM Data on equation coefficients, row by row
  220 DATA 5,2,1,4,4
  230 DATA 1,1,1,1,4
  240 DATA 1,2,3,4,6
  250 DATA 2,2,1,1,7
  260 DATA 1,2,3,4,5,6,7,8,9,0
  270 DATA -1,3,4,2,-5,1,1,1,1,0,0,12
 8000 REM Inverts NxN matrix in array A(1 to N,1 to N) into
      itself
 8010 REM Uses 0 column of A for calculations
 8020 DT=1:FOR L=1 TO N
 8030   Q=0:FOR K=1 TO N
 8040     Q=Q+A(L,K)*A(L,K):NEXT K
 8050   Q=SQR(Q):DT=DT*Q:NEXT L
 8060 DM=1:FOR L=1 TO N
 8070   A(0,L)=L:NEXT L
 8080 FOR L=1 TO N
 8090   P=0:M=L:FOR K= L TO N
 8100     IF (ABS(P)-ABS(A(L,K)))>=0 THEN GOTO 8120
 8110     M=K:P=A(L,K)
 8120     NEXT K
 8130   IF L=M THEN GOTO 8160
 8140   K=A(0,M):A(0,M)=A(0,L):A(0,L)=K
 8150   FOR K=1 TO N:Q=A(K,L):A(K,L)=A(K,M):A(K,M)=Q:NEXT K
 8160   A(L,L)=1:DM=DM*P
 8170   FOR M=1 TO N:A(L,M)=A(L,M)/P:NEXT M
 8180   FOR M=1 TO N
 8190     IF L=M THEN GOTO 8230
 8200     P=A(M,L):IF P=0 THEN GOTO 8230
 8210     A(M,L)=0
 8220     FOR K=1 TO N:A(M,K)=A(M,K)-P*A(L,K):NEXT K
 8230     NEXT M
 8240   NEXT L
 8250 FOR L=1 TO N
 8260   IF A(0,L)=L THEN GOTO 8330
 8270   M=L
 8280   M=M+1:IF A(0,M)=L THEN GOTO 8300
 8290   IF N>M THEN GOTO 8280
 8300   A(0,M)=A(0,L)
 8310   FOR K=1 TO N:P=A(L,K):A(L,K)=A(M,K):A(M,K)=P:NEXT K
 8320   A(0,L)=L
 8330   NEXT L
 8340 DM=ABS(DM):DT=DM/DT:RETURN
```

Figure 6.3   Listing for MATINV; subroutine from line 8000 inverts the $N \times N$ matrix in the array A(I,J) into itself; the first part of the program illustrates the application to solving simultaneous linear equations

(a)

(b)

(c)

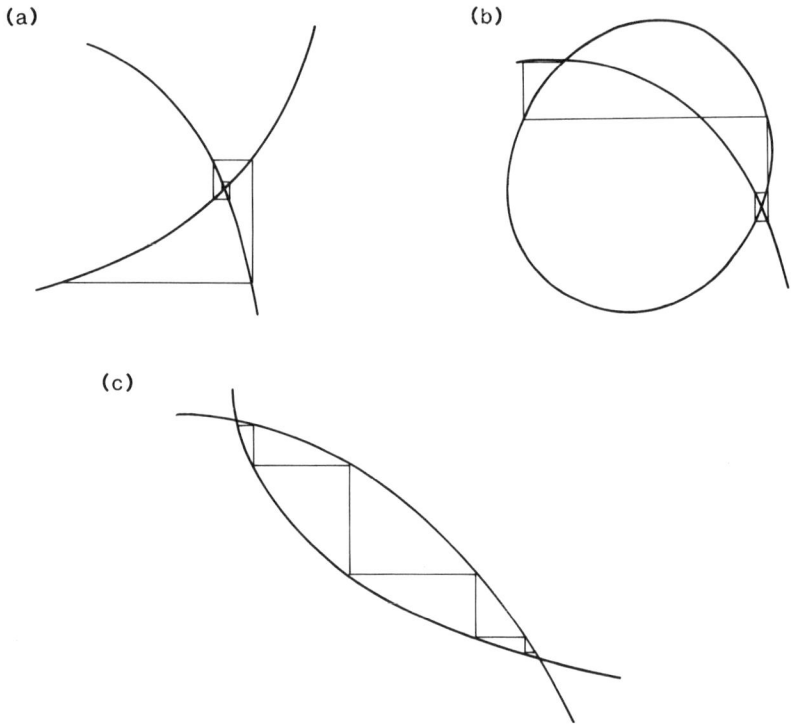

Figure 6.4 Graphical illustration of possible convergence or divergence paths for two non-linear simultaneous equations represented by curves; zigzag iteration paths can be followed in either direction according to choice of which unknown is derived from which equation

and similarly for V. It is important to ensure that the before and after values of H are at equal distances $(\Delta M/2)$ either side of the position at which the differential is being evaluated. If it is not possible to ensure that this condition is met, then there is some tendency for unintended effects to be introduced.

Since the difference equation is an approximation, a solution based on it will be most accurate when the increments of M, and consequently of V and H, are as small as possible. Without entering into a formal analysis, it is advisable to allow the increment to be changed in runs of your final program, so that you may compare the effect on the forecast values. You may like to experiment in this way with the program GROWVEG to determine what you feel is a reasonable compromise between precision and speed. Equations (3.20) and (3.21) are a pair of equations of very slightly greater complexity, and once more the transition from the differential to the computable difference equations is entirely intuitive, provided that care is taken to estimate the differential over appropriate spans.

The program SLOPEN in Chapter 4 may be used as an illustration of a partial differential equation. This necessarily involves at least two independent variables. SLOPEN forecasts slope evolution under transport limited hillslope sediment processes, such as creep and wash. Figure 6.5 is a flow diagram for the model, in which the slope profile is represented by a sequence of similar stores (as opposed to the one in STORFLO and the two in GROWVEG). Here the size of the space (x) increment determines the number of stores (N) down the length of the slope. If the stores are numbered 0 to N, then we may consider a storage equation and process laws for the Rth store, and this case may be generalized to all of the stores, except for the special conditions at the two boundaries (0 and N). The storage or mass balance for the store has the difference form, repeated here from equation (4.5):

Increase in storage during ΔT = Sediment inflow − Outflow

$$\Delta_t z(R)\, \Delta x = (S_{IN} - S_{OUT})\, \Delta t = -\Delta S \Delta t \qquad (4.5)$$

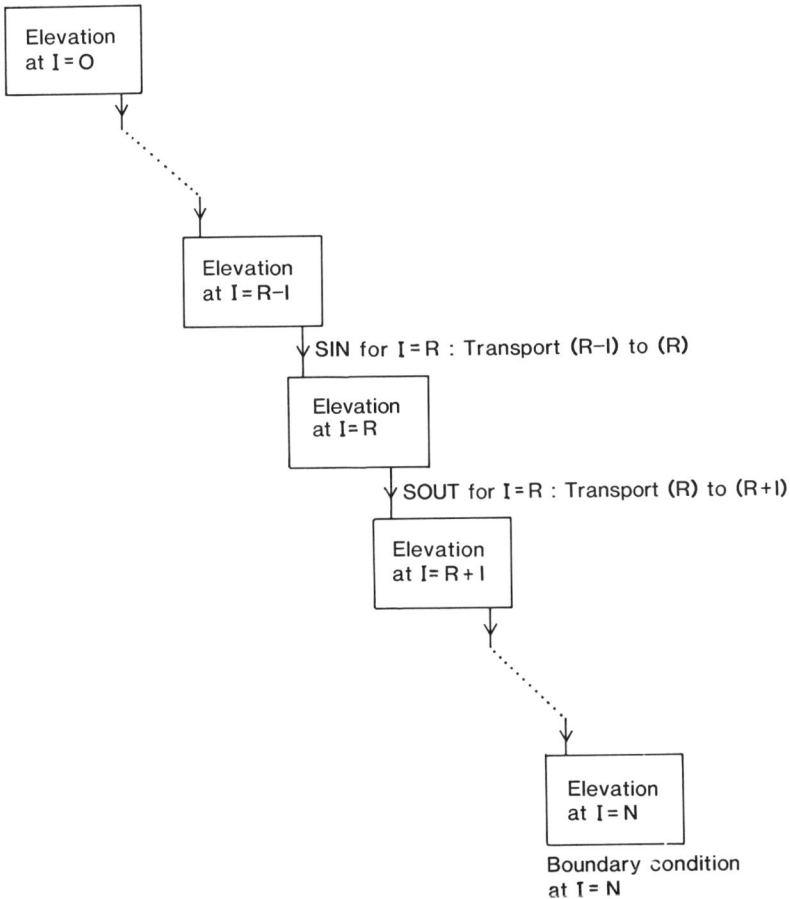

Figure 6.5 Systems diagram for SLOPEN

The change in storage on the left-hand side is the change of elevation over time at R, $\Delta_t z(R)$, which is spread over the width Δx of the storage cell. The relevant sediment inflow, S_{IN} is that from $R-1$ to R; and the outflow from R to $R+1$. These rates of flow are accumulated over the time increment, Δt.

When we look at the form of the sediment transport terms, they will be seen to contain terms which depend on slope gradient, which is the downslope change in storage z in the downslope direction, per unit distance, which may be expressed as $-\Delta_x z/\Delta x$, the negative sign coming from the convention of decreasing elevation with increasing distance, indicated in Figure 6.5. This gradient should be measured between the two stores, to represent the flow between them, so that the available elevation points conveniently provide the correct span of measurement. Since we are differencing a sediment term which contains gradient in it, equation (4.5) represents a partial differential equation in x and t, with first and second differentials in it. In other words, an intuitive approach can provide an adequate solution to a range of both ordinary and partial differential equations. The only additional problem with the partial equation is to choose suitable values of Δx and Δt to make the best compromise between accuracy and speed. For most partial differential equations, the time taken to perform each iteration begins to be noticeable, and may range from a second up to several minutes, according to the micro used and the complexity of the model. It will be found through experimentation that if too large increments are chosen, the simulation will overshoot its correct value, and set up oscillations in the solution which are entirely an artefact of the program. This condition must be avoided, usually through reducing the time step. If the distance increment is also changed, then the same degree of stability can be achieved by keeping the product $(\Delta x)^p (\Delta t)^q$ constant, where p is the order of the differentials with respect to x, and q the order with respect to t. For the SLOPEN example, $p=2$ and $q=1$. It may be inferred that the model will run faster for a given degree of precision with a small number of points downslope (large Δx) and a short time increment; since the run time is roughly proportional to the product $1/(\Delta x\ \Delta t)$. Alternatively a variable time step may be used, as illustrated in Section 4.4.2 for SLOPEN.

6.3 FLOW DIAGRAMS

Once the systems relationships and/or sets of equations have been defined appropriately as discussed above, the logical operations must be placed into a sequence so that they can be performed within the computer in a definite order. In many simple cases this process is transparent, but systems of equations are often mutually interlinked, and it is then incorrect to update the causal variables before all the relevant flows have been calculated. The safest method is to set up separate variables for the various net changes to each system variable over a single time step; and to update all the variables at the end of the time step. This can in many cases be avoided by skilful sequencing, but requires care. An alternative is to define an array of values, one for each time increment, but this can quickly overstretch available memory on a microcomputer.

Thus in computing changing vegetation and organic soil, following equations (6.1) or (6.8), it is incorrect to recalculate V or H before the flows have all been defined. If V is updated from the first equation, then calculation of the equation for H must avoid recalculating leaf fall on the basis of an updated V. The method actually used in GROWVEG is to calculate leaf fall and decomposition first, and then to update V and H without changing these flows. In a more complicated example, like SLOPEN, it is good practice to set up a rate of change matrix, $DZ(R)$, and to store all changes in it before updating any elevations in the $Z(R)$ matrix.

The sequence of operations can be clearly expressed in a flow diagram, which shows the sequence of operations, together with decision points, at which the sequence can branch on the basis of a logical choice. There have been some examples in previous chapters, including Figures 1.6 (GRAPH) and 3.4 (STORFLO). The simplest possible flow diagram consists of a straight chain of sequential operations, without loops or branches. Figure 3.4 shows slightly more variety, with a main loop for time, and a decision about presence or absence of overland flow. In detail the STORFLO program shows many more loops associated with checking input data values. Figure 6.6 shows the slightly more involved looping in the TIMAREA program in Chapter 2, again ignoring input and output routines. The flow diagram has been drawn in a more compact and stylized form. Loops have been identified by 'begin Time loop' and 'end Time loop', etc., with an arrow returning from the 'end' to the 'begin' to indicate a possible branch from the straight downflow of the program. Other decisions are shown by a question mark and an arrow for the alternative route, labelled with 'Yes' or 'No'. In some cases, though not on this diagram, the program needs to transfer control unconditionally. It is good practice to minimize the use of such transfers,

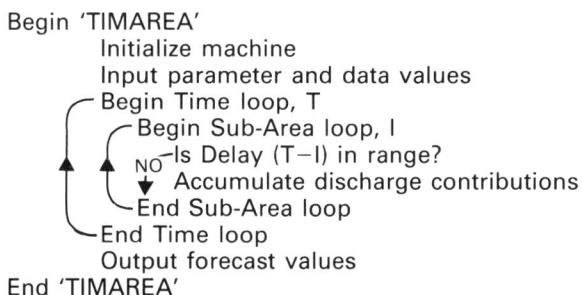

```
Begin 'TIMAREA'
        Initialize machine
        Input parameter and data values
        Begin Time loop, T
            Begin Sub-Area loop, I
              NO─Is Delay (T−I) in range?
                  Accumulate discharge contributions
            End Sub-Area loop
        End Time loop
        Output forecast values
End 'TIMAREA'
```

Figure 6.6   Flow diagram for central routines of TIMAREA

which are shown in the diagram as 'Jump', with an arrow to show the destination.

The process of flow diagram development will be illustrated using the slope evolution program SLOPEN from Chapter 4 as an example. The main alternative approaches to development are called *top down* and *bottom up* programming. In the top down approach, we begin with the grand design, attempting to identify main blocks of the routine which can be done in sequence, with as little branching as possible at this stage. Each block can then be broken down into smaller blocks, within each of which the detail can be expanded; and so on until all the detail has been worked out. The alternative, bottom up approach begins with the details and builds them up into a complete routine. This method has the advantage that the cross references between detailed units are likely to be more explicit. Against this must be set the severe risk of not seeing the wood for the trees. In practice the top down approach is strongly recommended as the main thrust of program development, with some of the details and cross referencing worked out on a bottom up basis. This procedure is analogous to the way in which a report is written or a map drawn. One begins with a series of headings, or a frame and a scale; and fills in the sections at growing levels of detail. At the end it is commonly necessary to check through the details to ensure proper cross referencing and consistency. In all cases it is clear that a part of the initial planning must include some provisional rules to maintain consistency between sections. This may be a set of map conventional signs, a standard system for footnotes or citations in a report, or a series of standard input/output routines and variable names for a program. We will ignore this aspect of planning at this stage, but it becomes very important in converting a flow diagram into a working program.

Figure 6.7(a) shows perhaps the very first stage in planning the slope model SLOPEN. At this level the steps each represent a major part of the overall routine, and at this level many programs look alike. The stages have been defined to avoid any looping. Each of the operations shown in Figure 6.7(a) can be expanded into a separate flow diagram. Figure 6.7(b) shows how this has been done for the main section, entitled 'Run model' in Figure 6.7(a). The whole model runs over time, represented by the outer loop. Within it, there are two inner loops running down the length of the slope in the X direction, the first of which calculates all the erosion rates in the array DZ, and the second of which updates the elevations, to avoid changing them prematurely. In Figure 6.7(c) additional detail has been added for the routine 'Calculate erosion rates', which includes tests for the special conditions at the top and bottom of the slope. It is at this level of detail that an element of bottom up development may become necessary. It is assumed in Figure 6.7(c) that the basal boundary condition will provide a rate of lowering Δz directly, so that in this case the routine jumps the stages which otherwise calculate Δz. The top boundary condition is, however, included on the assumption that it will provide an input sediment transport, S_{IN}, so that Δz

(a)   Begin 'SLOPEN'
                Initialize machine
                Input parameters and data
                Generate initial slope profile
                Run model
        End 'SLOPEN'

(b)   Begin 'Run Model'
                Begin Time loop, T
                    Begin Downslope loop, X
                        Calculate erosion rates
                    End Downslope loop
                    Begin Downslope loop, X
                        Update elevation at X
                    End Downslope loop
                    Output forecast values
                End Time loop
        End 'Run Model'

(c)   Begin 'Calculate erosion rates'
                Is X=Base X?
                Basal condition for DZ
                                                NO
                Jump
                Calculate S_{OUT}
                Is X=Top X?   NO
                Top Boundary condition for S_{IN}
                DZ=(S_{OUT}-S_{IN})/DX
                S_{IN}-S_{OUT}
        End 'Calculate erosion rates'

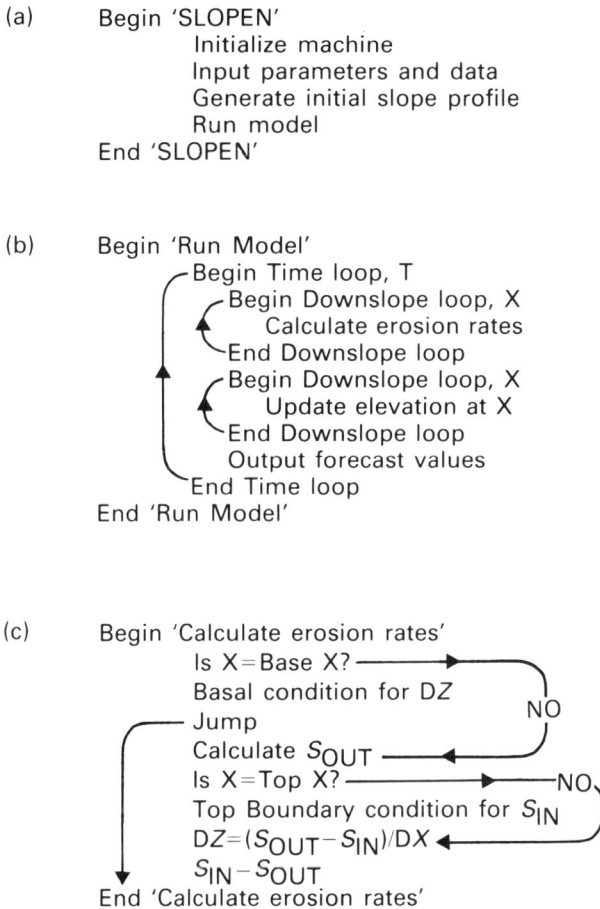

Figure 6.7 Flow diagrams for SLOPEN (a) the overall program outline; (b) the routine 'Run model' within (a); (c) the routine 'Calculate erosion rates' within (b)

is calculated from it in the normal way. These assumptions depend on the nature of the boundaries, which might therefore influence the program flow in detail.

6.3.1 Flow diagrams for logical and relational routines

We will now turn from flow diagrams generated from sets of equations to those where the routine is concerned with finding and manipulating the relationships between data. These routines form an important component of some models, particularly where data need to be sorted and/or the relationships between data need to be examined. Perhaps the most basic process is that of sorting a series of numerical or alphabetic values into ascending or descending order. There are many algorithms for sorting data. Figure 6.8 illustrates two of them for rearranging the values of a numerical array $A(1$ to $N)$ into descending order. It can be seen that the algorithms differ greatly in concept. 'Sort1' looks for the highest value and places it in first place. It then looks through the remainder of the list, and places the next highest value in second place, and so on. This is a simply intuitive

```
         Begin 'SORT1'
(a)            ┌──── Begin J loop (1 to N)
              │        MAX = -1e9
              │      ┌ Begin I loop (J to N)
              │     │  ┌ Is A (I)> =MAX?
              │     │ NO   MAX=A(I): K=1
              │      └ End I loop
              │        A(K)=A(J): A(J)=MAX
              └──── End J Loop
         End 'SORT1'


(b)      Begin 'SORT2'
              Set up array B (1.5*N,1)
              K=0: L=1: M=N: KM=1: B(0,0)=1: B(0,1)=N
         ┌──── I=M-L+1
         │     Is I=1? ──────── YES ──────────→
         │    ┌ Is I>2?
         │    │ Is A(L)> =A(M)? ──── YES ──────→
         │ YES│ Swap A(L) with A(M)
         │    │ Jump ─────────────────────────→
         │     R=A(int( (L+M)/2) ): L=1: M=N
         │    ┌→ Is A (L)< =R? ◄─┐
         │    │ YES L=L+1         │
         │    │   Jump ───────────┘
         │ YES│ → Is A(M)> =R? ◄─┐
         │    │ YES M=M-1         │
         │    │   Jump ───────────┘
         │     Is L>M?
         │ NO  Swap A(L) with A(M)
         │     L=L+1: M=M-1
         │   ←─ L=L< =M?
         │     B(KM,0)=1: B(KM,1)=B(K,1): KM=KM+1
         │     B(KM,0)=B(K,0): B(KM,1)=N: KM=KM+1
         │     K=K+1 ──────────────────────────
         │     L=B(K,0): M=B(K,1)
         └ NO ─ Is K=KM?
         End 'SORT2'

         Begin 'SWAP A(L) with A(M)'
              T=A(L): A(L)=A(M): A(M)=T
         End 'Swap A(L) with A(M)'
```

Figure 6.8 Flow diagrams for two methods for sorting numerical values in descending order (a) 'Sort1': an intuitive approach; (b) 'Sort2': an algorithm based on the Quick Sort method, which is faster for more than about 50 items

procedure. 'Sort 2' is a much less intuitive method, but one which is quicker for more than about 50 values. It takes an arbitrary value R from the middle of the list, and partitions the array, with all values more than R in the first part, and all values less than or equal in the second part. It then repeats this operation with each partition, and continues subdividing until each part has one or two values in it. If there are two values, they may need exchanging: if one, there is nothing more to do. This method may also be seen to provide an effective way of sorting the array, as well as illustrating that there are often many possible algorithms for a given problem, and that the most obvious is not necessarily the best.

A second example of a relational routine is the subroutine STR_AN listed in Figure 5.8. This routine converts a binary string of M 1s and M 0s, representing the topology of a network of magnitude M, into a string which gives the Horton–Strahler order of each link; and a bifurcation matrix (a_{ij}) which shows how many links of each order i meet links of all orders j. Figure 5.11 is a sample of part of the output. Figure 5.7 shows how the network is scanned, in a clockwise direction, to describe the topology of each node when it is first visited (in an upstream direction). After an initial '1' for the outlet, nodes are denoted by '1' for interior nodes (junctions) or '0' for exterior nodes (stream heads). Figure 6.9(a) shows the topology of the network analysed in Figure 5.12, although the lengths are not correctly represented.

The algorithm used to analyse the string breaks down the network into the simplest possible structure, namely the sequence '100' in the binary string, which represents a stream dividing into two exterior links. If the two exterior links are removed, then the structure is replaced by a single exterior link, represented by a single '0'. This process may be repeated indefinitely, until the network consists of a single link and the outlet, so that the whole network has been reduced to the string '10'. In order to keep track of where we are in the network, it is better not to shorten the string, but to replace the '100' by '022'. In subsequent scans of the string, 2s are simply treated as though they were not there.

At the same time, the Horton–Strahler rules of stream ordering may be applied. These are that all fingertip tributaries have order 1; and that at a junction the order of the combined stream is equal to that of the higher tributary unless both are equal, in which case the order is increased by 1. To apply these rules, we can first identify all fingertip tributaries, with zeros in the original string, as of order 1. In the reduced binary string (with 2s in it as described above), as we locate successive '100' structures, we must already know the order of the '0' links, so that the rules give the order of each '1' as we change it to a '0', ready to be used again at the next stage.

These values are held in three strings. M$ holds the original binary string, and is left alone. Q$ is initially set equal to M$, but is progressively modified as '100's are converted to '022's. A$ is used for the orders of each link, initially with 1s for the first order streams (0s in M$) and asterisks elsewhere. As each pair of nodes is eliminated, the number of links meeting with each

140

(a) Topology of the network shown in figure 5.12.

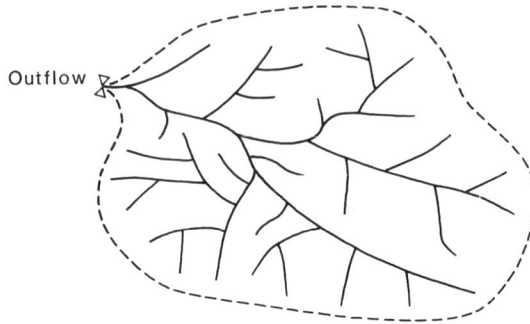

Outflow

(b) Overall flow diagram.
Begin 'STR_AN'
        Check for valid binary string M$
        Q$=M$ (Modified binary string)
        Transfer first order links to A$ (Order string)
        Search for 100
        Is 100 found? ————▶——— NO ┐
        Apply ordering rules
        Jump
End 'STR_AN' ————————◀———

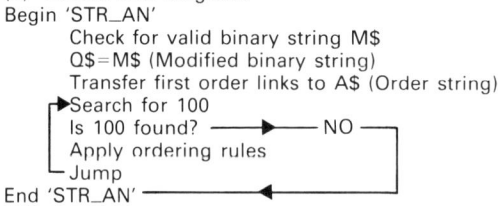

(c) The routine 'Search for 100' within (b)

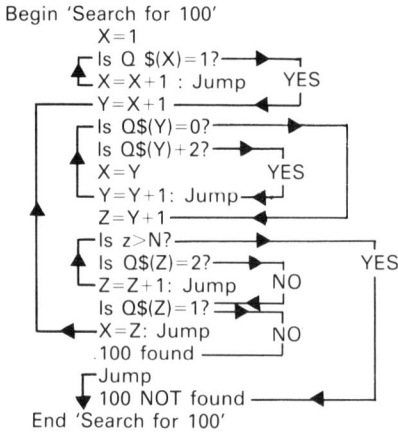

Begin 'Search for 100'
        X=1
        Is Q $(X)=1?———▶—
        X=X+1 : Jump   YES
        Y=X+1 —————◀——
        Is Q$(Y)=0?————▶
        Is Q$(Y)+2?——▶—
        X=Y   YES
        Y=Y+1: Jump—◀┘
        Z=Y+1 ——————◀—
        Is z>N?————▶
        Is Q$(Z)=2?——▶—   YES
        Z=Z+1: Jump   NO
        Is Q$(Z)=1?▶◀—
        X=Z: Jump   NO
        100 found ————┘
        Jump
        100 NOT found ———◀——
End 'Search for 100'

Figure 6.9   Algorithm used in STR_AN for analysing a sequence defining network topology to provide Horton–Strahler orders for each link and stream (a) topology of the network shown in Figure 5.12; (b) overall flow diagram; (c) the routine 'Search for 100' within (b)

order is accumulated in the 10×10 array BF. Figure 6.9(b) gives an overall flow diagram, with repeated searches for '100' structures; together with the details of this search procedure through Q\$ (Figure 6.9(c)). In fact the algorithm shown, and used in the program, may be modified by including the 'Search for 100' routine within a FOR loop, as we know that exactly $M-1$ searches will be needed to reduce the string to '10'.

It may be seen that a variety of methods, and considerable ingenuity backed by experience, is needed to make relational algorithms, like the examples here, work effectively. There are usually minor elements of this type in most programs, so that it is worth studying examples to improve your repertoire of possible analogies. The reader may like to follow the logic and produce flow diagrams for the programs CHNET and FRACTAL in Chapter 5.

6.4 INPUT AND OUTPUT DESIGN

Your program must communicate with you as you develop it, and with its final end users. To meet this requirement, the next stage of model design is to choose appropriate forms of input and output, and to incorporate them in the flow diagram. For large volumes of data, disc or tape files are almost certainly the best forms for entering large amounts of data; and similar files together with printed output are best for getting data from the computer model. At the level we are working at in the models of this book, and with a microcomputer, it is perhaps most realistic to concentrate mainly on input and output through the monitor or TV screen, as is used for the majority of the programs in this book. Facilities exist within the PROTO subroutines for copying a text or graphics screen to the printer, so that this route provides at least one way to obtain a permanent copy of results. We will return to discuss methods for handling larger amounts of data at a later stage.

The screen has a rather limited capacity for either text or graphics. Most micros offer a screen of 40 columns and 24 to 25 rows of text. Many offer a higher resolution screen as well, commonly of up to 80 columns $\times$ 25 rows. We will concentrate for now on the former, since anything that can be done in 40 columns can be done more easily with 80 columns. The screen can hold about 1000 characters or 125 words. This means that only a very limited amount of material can be displayed at once. For example, a display limited to three-digit whole numbers can display up to eight columns of figures (with a gap each side of the longest figures). If we have figures like '+12.34', which need six columns plus two spaces, only five columns can be displayed. If we are unsure about the sizes of the largest figures, we may be even more restricted. Thus any sort of table of data needs to be very carefully designed, with tight control on the location and format of printing on the screen. Where space is very tight, it may be necessary to format numbers very carefully, as in the subroutine at lines 930 to 970 of 'LINEAR' (Figure 4.8). There are also obvious restrictions on the length of row or column headings.

In the corresponding minimum graphics mode, resolution is likely to be about 256×200 plottable points or pixcels, although high resolution modes may roughly double the horizontal value. Once more, there are restrictions on the types of graphic design which are easily accommodated, so that it is best to rely mainly on simple line graphs, or histograms. Text labels can usually be added, but should be brief, and carefully located. Colour is generally, although not invariably available, and can help to clarify these diagrams. Some examples of the kinds of output readily available can be seen from the example models listed in previous chapters. It should be remembered that most current micros exceed the minimum specification used in at least some respects, so that programs tailored to any particular machine can be considerably less restricted in their output design. Similar restrictions of size apply to input, although it is rarely so serious because it is not usually necessary to see all of the input values at once, whereas it is useful to survey large blocks of output data together.

As well as the problem of screen design, care should be taken with the timing and quantity of input and output. In using a micro interactively, it is important that the machine appears to be responding to the user, neither appearing to be doing nothing at all nor swamping him or her with excessive demands or excessive output. The difficulty here is that a good deal more output is usually required during program development than for its final use. This usually affects the frequency of output, but also its level of detail and the number of intermediate results, etc. The development program may therefore include output routines which will not be wanted in the final version. The best advice here is to include variables which control the frequency of output, so that it can be reduced eventually; and to include additional routines which are concerned solely with the output of intermediate results, and which can ultimately be neatly cut out without influencing other aspects of program execution. As a model runs, there are sometimes routines which take several seconds to minutes to perform. The user should not be left in any doubt about whether his computer is working for him, or has broken down or gone into some kind of infinite loop. This reassurance can usually be achieved by printing out some value, often a time or a loop variable value, in an out of the way corner of the screen, so that the user knows that something is going on, and the program developer knows exactly what stage the program has reached.

For the main output of a simulation model, it is often a matter of individual preference whether to show progress through a graph or through text output. A graph often gives a better 'feel' of what is going on, and so may be preferable. On the other hand, it is imprecise and will generally need a text printout to back it up in the end. Unless computer memory is limited, it is usually possible to provide both forms of output from arrays of values. An array also has the advantage that maximum and minimum values of the output for each run can be calculated, and graphical (and to a less extent textual) output scaled so that the results effectively fill the screen, instead of forming an indecipherable squiggle in one corner. Where there is insufficient memory to store an output array, the best solution may be to offer the user a choice, so that a model run can, if required, be replicated to give the alternative style of output.

Input problems relate mainly to the number and meaning of the parameters entered, rather than to issues of screen design. At least two extreme styles of input may each be acceptable in suitable cases. In the first style, all input data values are included in the program (usually in DATA statements), so that the program runs without user input. Change of data values is then a matter of program editing. This is simple and effective in a development program, but is unsatisfactory for a non-expert user. The alternative extreme style is to ask the user to input every single piece of data needed. This method may overload the knowledge of the inexpert, and the patience of the expert if more than a very few values are needed. It may be backed up by various degrees of prompting, as can be seen from our example programs. Prompts may for example give the acceptable range of values, and check that input values lie between them. It may also offer a suggested value, which the user can amend or accept. As a general rule, the ideal is perhaps to offer the user a restricted number of choices, with as much context as possible to allow them to be made rationally; and to offer the option of editing a more complete data set.

For large input or output data sets, the use of a file on disc or tape is the most flexible. Such files should be set up so that they can be printed directly, to give a comprehensible document. You will also need facilities for editing the input files of data, either within the model program itself or as a distinct program, associated with it. Examples of file use and editing may be seen in RUN1 (Figure 2.11) and LINEAR (Figure 4.8).

The final stage in designing input or output is to include it within the overall flow diagram for the model. Figure 1.6 is an example of the flow chart for a simple program to draw a graph of a function. In this case successive values are calculated and drawn directly, but the same routine is equally effective for plotting values taken from a data array, either as its values are produced, or as a subsequent operation. Figure 6.10 gives a flow diagram for INDATA, which is listed in Figure 1.2, and is a typical routine for entering a sequence of values. Variations are needed to enter a sequence of values with the same title, but at a different numerical value, like a sequence of rainfall amounts in successive hours (as in TIMAREA and STORFLO), in that there is no need to read in a new title and suggested values for each. In such a case, there is also a choice about whether to enter successive rainfalls (or whatever) together at the start of the program, or whether to enter them one at a time, as the model predicts the associated runoff (or whatever is forecast).

Similarly, in designing the sequence of outputs, there is the need to reassure the user that the machine is working correctly, and to give enough but not too much data. In a time sequence with many iterations, we may

The routine inputs a series of values, cued by suggested values and range limits. A listing is given in figure 1.2.

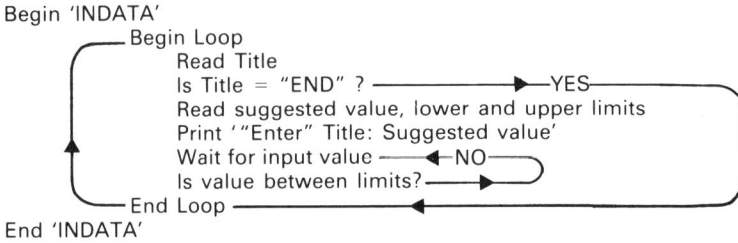

```
Begin 'INDATA'
        ┌── Begin Loop
        │       Read Title
        │       Is Title = "END" ? ──────────►YES──────────┐
        │       Read suggested value, lower and upper limits │
        │       Print ' "Enter" Title: Suggested value'      │
        │       Wait for input value ──◄─NO──┐               │
        │       Is value between limits?─────►┘              │
        └── End Loop ──────────────◄──────────────────────┘
End 'INDATA'
```

Figure 6.10   Flow diagram for INDATA

look for a compromise in which, say, a one-line output records the passing of each iteration and one or two key values, but a whole screen full of data is supplied either at a fixed number of iterations, or when the model has evolved to relevant stages measured in performance rather than time terms. In this way, the user has time to absorb the screen data between renewals of it, and may be able to request a printout by pressing a key.

Input and output are seen as important aspects of an effective model, and can make a great deal of difference to how pleasant a program is to work with. Nevertheless, our concern here is with development programs rather than highly finished and foolproof pieces of commercial software, so that it is also important not to spend too disproportionate a time in embellishing the program in this way. We should not lose sight of the original aims of each model, and its input/output facilities should generally be designed with function rather than beauty in mind!

6.5   WRITING A BASIC PROGRAM

This section is related exclusively to BASIC programming. It concentrates on the minimal set of commands common to almost all versions of the BASIC language, which are listed in Appendix A. Most versions of BASIC are less restrictive in various ways, and it is generally efficient to make use of the special features of your particular BASIC dialect, although at some cost to transferability. In particular it is worth using 'procedures' or other devices for effective parameter passing to avoid the need for such strict constraints on variable names in using subroutines (or their equivalents). It is also very well worth while to make direct use of individual graphics commands. The guidelines below assume that procedures (as for the BBC and QL, or for 'BetterBasic' on the IBM PC) are not available, and that input and output are largely handled through the machine specific subroutines of PROTO given in Appendix B.

The aim of successful programming is to write programs which somebody else can read and understand without difficulty; and which perform efficiently.

Comprehensibility can be achieved in a number of ways, but most rely on clear documentation. The most direct method is to include a large number of REM (remark) statements in your program text, but they take up some computer memory and slow down program execution somewhat, especially if included within frequently used loops, etc. There are advantages therefore in using a small number of REMs in support of separate documentation. This may take the form of a program listing in the left-hand 40 columns of a page, with written commentary beside it on the right-hand half of the page. A better alternative is to use separate documentation which contains descriptions of the program's purpose; discussion of input values and their meanings; samples of output, preferably as it will appear on the screen; and flow diagrams at one or more levels to explain difficult points. The program itself should then be arranged in a way which can readily be related to the flow diagram, with a few REMs and blank lines to clarify the connection. Most of the program listings here are relatively modest in their use of REMs, although the number in the main loop (lines 130 to 290) of KINWAVE in Chapter 5 is perhaps excessive.

Efficient program execution relies mainly on the development of a good algorithm and flow diagram. At the programming stage, it is important to make best use of the various control statements which carry out loops, decision branches and jumps. In addition, programs can be made briefer, although not usually faster running, by effective use of subroutines (including those in PROTO if used). The following notes are intended as a guide to the sensible use of control statements.

6.5.1   GOTO line number

Unconditional jumps generally confuse the flow of the program, and should be avoided when possible. There are always some occasions when GOTO is convenient and a few when it must be used, but excessive use is very confusing, as may be seen from this example, where its use is extreme and totally unnecessary.

```
10   RESTORE: READ X,Y,Z: GOTO 30
20   INPUT C:GOTO 40
30   INPUT A:GOTO 50
40   PRINT A*X,B*Y,C*Z:GOTO 30
50   INPUT B:GOTO 20
```

On the other hand, the GOTOs in Figure 6.2 could not be avoided without a much more repetitive listing.

6.5.2   IF logical expression THEN statement

This allows the evaluation of any logical statement as true or false. If it is true then the statement(s) following THEN on the same program line are executed. Otherwise the program moves on to the next consecutive line number after the IF statement. In many cases the statement after THEN will be a GOTO, but careful choice of the logical expression can minimize this. If a large number of additional statements need to be executed if the logical expression is true, the statement following THEN can make use of subroutines (see GOSUB below) to clarify the flow of the program.

6.5.3 FOR variable = beginning TO end (STEP increment)

Where one needs to perform an operation a previously known number of times, then the FOR loop achieves it neatly and efficiently. All statements between this statement and the ending statement 'NEXT variable' will be performed for each valid value of the variable, and usually at least once even if the end value for the variable is less than the beginning value. This form of loop should, however, be used with care if you want to get out of the loop before reaching the end value for the variable. It is good practice to ensure that the variable reaches its end value before leaving the loop; and not to use more than one NEXT with each FOR loop. One way to get out neatly is shown below, using the variable J to show the number of strings actually read before reaching the terminating "END":

```
100  J=N: FOR I=1 TO N
110  READ A$:IF A$="END" THEN J=I−1:I=N:GOTO 200
 .
 .
 .
200  NEXT I
```

6.5.4 Other types of loop using IF and GOTO

Although it is usually more efficient in BASIC to use FOR. . .NEXT loops whenever possible, there are occasions when other methods of looping are clearer. The FOR. . .NEXT loop is exactly equivalent in logic to:

```
 90  I=1
100  . . .
 .
 .
 .
200  I=I+1: IF I<=N THEN GOTO 100
```

This form of loop is normally appreciably slower in execution than the FOR. . .NEXT loop. For example, on the BBC 'B' micro, an empty FOR I=1 to 10000 loop takes about 7 ms, whereas the equivalent IF. . .GOTO loop takes about 30 ms. This saving in time becomes negligible when there are many statements to perform in the loop, so that this need not be an overriding consideration. The IF. . .GOTO loop may be preferable where the number of loop iterations is unknown (although you can simply put in a very large number), where the condition for exiting the loop does not depend on the loop variable, or where the loop need not necessarily be performed at all. In all cases, however, suitable decision statements allow the FOR. . .NEXT loop to be used. For example in the last of these three cases:

```
 90  IF M>N THEN GOTO 210
100  FOR I=M TO N
  ,
  ,
  ,
200  NEXT I
210  . . .
```

A number of examples of IF. . .GOTO loops may be seen in the listing of STR_AN in Chapter 5, where they may be seen to be convenient, representing the flow diagram (Figure 6.9(b)), even though their use is not absolutely necessary.

6.5.5 The use of flags

In a number of cases, it is helpful to make use of flag variables which have no meaning outside the program, to keep track of the state of various aspects of the program. For example the PROTO routines use a series of flags to note whether the printer is to be used or not, whether the graphics screen is in use, and/or has axes drawn on it, etc. Flags may also be useful to note whether you have completed a FOR. . .NEXT loop normally, or have jumped out prematurely. The following program fragment shows the use of a flag to run through a loop twice, first counting the number of items in a data list, and the second time assigning them to an array, Q$ of the correct size.

```
 90  F=1:N=1000
100  RESTORE:FOR I=1 TO N
110  IF F=2 THEN READ Q$(I):GOTO 130
120  READ A$: IF A$="END" THEN N=I-1:I=1000
130  NEXT I
140  F+F+1:IF F=2 THEN DIM Q$(N):GOTO 100
150  . . . .
```

6.5.6 Subroutines (GOSUB line number)

The effective use of subroutines (or procedures if available) is an important element of good programming. They are used mainly to avoid duplicating sections of program which would otherwise need to be repeated, but also have a useful role to play in improving the clarity of the program flow and in allowing the program to be used in a modular way.

Most of the program listings illustrate the repeated calling of subroutines 12000 or 12500 in PROTO to input a series of data values from the keyboard. KINWAVE (Figure 5.5) illustrates the use of subroutines nested inside one another. The main loop calls subroutine 700 to update the position of stones on the screen. In turn it calls subroutine 850 which calculates first the old and then the new position of the stone to be moved. This nesting can be taken further where appropriate.

As with FOR. . .NEXT loops, it is important to come out of a subroutine correctly. The program returns from each subroutine when it first encounters a RETURN statement. There can be more than one RETURN in a subroutine, the choice of exit depending on a decision branch or flag value. There is no way of identifying which RETURN statement should be paired with which GOSUB. Branching can therefore take place between subroutines provided that a very careful check is kept on the depth of nesting, to ensure that exactly the same number of GOSUBs and RETURNs are met with in any possible flow of program execution. Failure to meet this condition will generally lead to errors and/or unpredictable results during program execution.

The program SLOPEN in Chapter 4 and Figure 6.7 may be used to illustrate the use of subroutines to clarify the flow of the program and to give a modular structure, even though each subroutine is called from only one point in the program. The routines to generate the initial slope form, to calculate the sediment transport at each point downslope, and for boundary conditions at top and bottom of the slope, are each in a separate subroutine, placed at the end of the program listing. The main flow of the program can, as a result, be more readily identified with its flow diagram in Figure 6.7, each subroutine being described by a brief REM in the main part of the program. To alter any of the main aspects of slope development, the relevant subroutine can be modified without interfering with the overall program layout.

In detail, interfacing between a main program and its subroutines requires some care. It is vital to keep a check of what variable names are used in the subroutine, both internally and to communicate with the main program. A fairly safe way to avoid duplication is to use a definite pattern of variable names. For example, the main program might use single letter variable names only, both for internal use and to pass values to and from its subroutines. Each subroutine might then internally use two letter variable names, the second letter being different for each subroutine. Some such scheme avoids the possibility of mistakenly changing the value of a variable. Alternatively, a careful check needs to be made on which variable names may be changed within each subroutine. The most frequent problem is perhaps the use of a loop variable like I or J in the subroutine, and forgetting that one is in the middle of a loop with the same variable name in the main program! Figure 6.11 illustrates the variable names used for input and output of values, and for internal use, within the program KINWAVE. Subroutine 700 calls subroutines, first 850 and then 16000. The internal variables TX and TY are calculatcd by SUB 700, and can then be used to pass values to SUB 850. In turn, SUB 850 returns values for R and C, which with A$ (generated in SUB 700) are passed to SUB 16000. The original input variables NX and NY have to be returned to the main program loop unchanged. This kind of analysis is not needed in most cases, but can be helpful when your program appears to be acting strangely.

Subroutine	Input	Internal	Output
		- - -V A R I A B L E S - - -	
GOSUB 700	X, Y NX, NY	TX, TY R, C, A$	None
GOSUB 850	TX, TY	None	R, C
GOSUB 16000 (in 'PROTO')	R, C, A$	Various (according to type of micro)	None

Figure 6.11 Example showing the use of variables for interfacing with subroutines; variable names for input, output and internal use are shown for subroutines in KINWAVE

6.5.7 ON index GOTO: ON index GOSUB

A final control structure worth noting is the use of an index or flag (which must take integer values of one or more) to direct the flow of program execution in one of several directions, either unconditionally (ON. . .GOTO) or to subroutines and returning (ON. . .GOSUB). In each case the GOTO/GOSUB is followed by a series of line numbers, to which execution moves according to the value of the index: to the first if index=1; to the second if index=2, and so on. This is a helpful device for offering choice to the user through some form of 'menu', as well as a way of branching in more than the two directions allowed by a single IF. . .THEN decision. In some cases a number of the line numbers may be the same, and in some case they may point to an ineffective line consisting solely of a RETURN. The following fragment gives an illustration.

```
100   PRINT "Enter your choice":PRINT:PRINT " 0 for no output"
110   PRINT " 1 for Text output":PRINT " 2 for graph output"
120   INPUT A: A=INT(A):IF A<0 OR A>2 THEN GOTO 120
130   IF A=0 THEN END
140   ON A GOSUB 15000,20000:REM Open appropriate screen
150   ON A GOSUB 500,1000
      .
      .
      .
500   REM Text output subroutine
1000  REM Graphics output subroutine
```

6.6 PROGRAM DEBUGGING

The final, and usually one of the most time consuming stages is to make the program work as you originally intended. There are three main parts of this process, each of which is likely to need repeating more than once. The first stage is to make the program work at all, the second to ensure that it does what you want, and the third to refine it.

In most forms of BASIC, errors of syntax are reported back to the user as soon as a faulty line is entered at the keyboard. This usually deals with lines where you have unequal numbers of left and right brackets; have used a semicolon where a comma is required; have misspelt a keyword; or other small but vital details. You will need to look at your computer manual to interpret the 'error messages' which come back to you on the screen, and learn their idiosyncracies.

When you have corrected the syntax of each individual line, it is possible to run the program. At this stage many more error messages appear, and the program stops at the first one. Table 6.1 lists some of the commonest errors, and we each tend to have our own 'favourites' which we are particularly prone to commit. With practice, we make less mistakes, but you should not expect perfection. When you find it impossible to see what has gone wrong, it sometimes helps to repeat the run, adding a print statement just before the point of failure, to display some of the suspect variables, and the index values of loops, arrays and 'ON GOTO/GOSUB' statements. The subtler errors often arise because of a mistake earlier in your program which, while not itself an error which stops execution, nevertheless provides a false value (a zero or too large an index, for example) which causes the error later. When desperation sets in, though not before, it is also helpful to ask someone else to look for your errors.

When a program finally works, one is apt to assume that it is doing exactly what was originally planned. In all but the simplest cases, this is groundless optimism. What you have to do at this stage is to test the program, in all its parts, with data values for which you know the correct answer already. If this is too daunting a task, then it may be preferable to tackle it piecemeal.

Table 6.1 Common run-time errors in BASIC programs (all apply to the BBC micro; errors which do not apply on some other micros are shown by an asterisk)

1	Division by a variable which has value zero
2*	Failure to dimension arrays before use; or redimensioning an array already in use
3*	Failing to assign a value to a variable before using it (array values are set to zero when dimensioned)
4	FOR without correctly matched NEXT
5	Allowing program to reach a subroutine which has not been called; avoid by putting END before first subroutine, or GOTO around the subroutine(s)
6	Not enough DATA values to match number of READs
7*	GOTO or GOSUB a non-existent line number
8	On X GOTO/GOSUB . . . with X taking value zero, or more than number of line numbers in list
9	Referring to array with index less than zero, or beyond dimensioned range

Write a line of program which provides relevant variable values and calls a subroutine. Check the values by hand for a range of cases until you are sure that it works in all possible cases: you may then assume that it functions correctly, and go on to the next subroutine. When all the subroutines work, then test the section of program which calls them, again trying to cover all possible cases. It is sometimes helpful to test the main loop with some or all dummy subroutines (consisting of RETURN only) to check its logic, although not of course the values it produces.

There is no real short cut to this testing stage, although you may be able to think of some simple cases, for example with many of the input parameters zero, which will allow some runs which are relatively easy to check. In the end, repeated running of the model provides this sort of checking, but it means that you are likely to go on finding bugs in your program for quite a while, as special or extreme cases crop up. At this stage, it may help you to work mainly with graphical output, which, if less precise than text, allows you to spot serious anomalies much more quickly.

As can be seen, this stage of thorough checking merges imperceptibly with the process of obtaining some useful results from the model, until a stage is reached at which you begin to have confidence in your model. From now on, you will begin to look at your output to see whether your original design was ideal. Changes you make at this stage increasingly represent refinements of the original model and program design; although in some cases they can lead to a complete redesign of the model! In the light of the output, you may wish to modify the original functional expressions or logical sequence of operations. You may also wish to improve the design of input or output, for example to a more compact form, with the option of fuller printout at infrequent intervals or under user control. One tends to develop preferences and a style of one's own in these matters, making use of the flexible screen layout for both text and graphics which microcomputers commonly offer.

You may also be able to make some saving at this stage in either the space which the program occupies in computer memory, or in the speed at which it runs. Table 6.2 lists a number of ways of making the program shorter. This can be done by cutting out unnecessary statements, notably REMs, and by saving on the amount of space needed to store variables, and especially arrays which are not really needed. Many, though not all of these savings are at the expense of readability of the program, so that they should only be used where the space gained can be put to good use. For

Table 6.2 Methods for saving memory space in BASIC programs (all apply to the BBC micro; methods which do not apply to some other micros are shown by an asterisk)

1* Omit all unnecessary spaces from program listing

2 Omit all unnecessary keywords; e.g. LET is usually optional and THEN GOTO can be replaced by THEN

3 Omit all REM statements

4 Assign frequently used constants as variables, e.g. K=454, and use the variable name when the constant is required; The same applies to string constants, e.g. Q$="+++++++++++"

5 Put frequently used sequences of instructions in a subroutine (or function); this can apply to print headings as well as calculations

6* Use integer variables (ending with %) wherever possible

7 Reuse the same array later on in the program if its previous contents are no longer required

8 Use short variable names for heavily used variables

9* Use multi-statement lines with several statements on each, separated by colons

10 Write IF. . .THEN statements economically, for example

```
100   IF X=Y THEN 120
110   X=Z
120   X=X+1
```

can be reduced to:

```
100   IF X<>Y THEN X=Z
110   X=X+1
```

11 Do not enter values into an array unless they are needed more than once

12 Use data files on disc/tape to replace long lists in DATA statements

Table 6.3   Methods for increasing program execution speed (all apply to the BBC micro; asterisked methods do not apply to some other micros)

1   Most of the space-saving methods in Table 6.2 will also increase speed

2   Avoid repeating calculations, particularly inside loops; for example
       INPUT X: FOR I=1 TO 10: PRINT I*SIN(X):NEXT I
    should be written as
       INPUT X: K=SIN(X):FOR I=1 TO 10:PRINT I*K: NEXT I

3   Use addition and multiplication rather than subtraction and division if
    possible, e.g. X=Y/5 should be rewritten as X=Y*0.2

4   Avoid mixed mode arithmetic, i.e. a mixture of integer and floating point
    variables

5   For nested loops, put the most frequently scanned loop innermost, for
    example
       100   FOR I=1 TO 500
       110   FOR J=1 TO 10
       120   A(I,J)=1
       130   NEXT J: NEXT I
    should be rewritten as:
       100   FOR J=1 TO 10
       110   FOR I=1 TO 500
       120   A(I,J)=1
       130   NEXT I: NEXT J

6   With long logical expressions, put the tests most likely to fail on the left

example the maximum size of data arrays may be increased usefully. Savings of space are in some cases at the expense of speed, and vice versa. Table 6.3 shows some of the ways in which program execution can be speeded up. In simulation programs, savings made in the most frequently used loops or subroutines have the greatest impact overall, so that it is worth using all possible devices to accelerate them where program speed is critical. As with space savings, gains in speed are not always needed for the comfortable use of a program. You can judge this for yourself from the examples in this book. Simulations over time, particularly those with a large number of stores like LINEAR or COLONY, can never run fast enough; whereas programs like INDEX, which spend most of their time waiting for input from the user, would show no advantage in running faster.

The impression you may get from the sequence of checking and rechecking recommended here, is that a program is never finished and perfect. That is true, but little more true than for most tasks. In every case, you must decide, for yourself or with the end user of the model, when the program is good enough. Beyond that point, further beautification only begins to waste the time you sought to save by using a computer in the first place. That way lies addiction, rather than effective use of your computer!

6.7 FURTHER READING

6.7.1 Numerical methods

Hornbeck, R. W. (1975) *Numerical Methods*, Quantum Publishers, New York.
Scraton, R. A. (1984) *Basic Numerical Methods*, Edward Arnold, London.

6.7.2 Modelling methods

Cross, M. and Moscardini, A. O. (1985) *Learning the Art of Mathematical Modelling*,
Ellis Harwood, Chichester.
Harte, J. (1985) *Consider a Spherical Cow*, William Kaufmann Inc., Los Altos,
California.

6.7.3 Learning BASIC (to be used in conjunction with the BASIC manual for your own micro)

Alcock, D. (1977) *Illustrating BASIC*, Cambridge University Press, Cambridge.
Clark, A. and Clark, J. (1984) *Self-instruct BASIC: A Practical Guide* (2nd edn),
Pitman, London.

6.7.4 Some standard methods and algorithms in BASIC

Mason, J. C. (1984) *BASIC Matrix Methods*, Butterworths, London.
Sharp, J. J. and Sawden, P. (1984) *BASIC Hydrology*, Butterworths, London.

Chapter 7

Model Calibration and Verification

7.1 STAGES IN MODEL DEVELOPMENT

In the last chapter we discussed how to construct a computer simulation model, and in particular, we considered how to structure the model. This entailed identifying the process elements, selecting mathematical equations or algorithms which best describe each process, and converting these algorithms to a computer program. Of course, we cannot even begin to design our model without knowing something of the real system being replicated. Most models depend on one or several types of inputs (e.g. solar radiation, rainfall). The structure of the model is described using two types of parameter: physical parameters provide information on the physical structure of the system (e.g. cliff height, basin area, soil depth); process parameters determine the order of magnitude of processes which affect the movement and distribution of the input (e.g. infiltration rate, stomatal resistance, weathering rate). The input, together with the physical characteristics and the process parameters, controls the response of the model at any one time. In many cases it is likely that we shall have observed the response of the system, giving us something to compare against the model's prediction, or output (although in some forecasting applications, this may not be the case).

It is necessary to quantify the process parameters in order to define the mathematical equations which make up the model; this action is known as calibration. Where the process parameters bear a physical resemblance to the actual attributes of the real system, calibration can be achieved by field measurement. Where this is not the case, they must be evaluated by a procedure known as optimization; in this case, the process parameters are more correctly known as optimized parameters. At this stage of model development, the model may be run using imaginary parameter values in order to investigate the 'sensitivity' of each parameter. This is described in more detail in the next section, and it is sufficient to note here that a 'sensitive' parameter is one which has a major effect on the model output, whilst an 'insensitive' parameter has much less effect. Thus, if the value of a sensitive parameter is changed slightly, it will have a much larger effect on the model prediction than if the value of an insensitive parameter is altered by an equivalent amount. It is important to measure (or estimate) the value of a sensitive parameter as accurately as possible; there is a clear link between the quality of data used to calibrate the model and the accuracy of the model's forecasts. Sensitivity analysis helps greatly in indicating where we must concentrate our efforts when calibrating a model.

The next consideration in model development is to assess model accuracy; this is normally achieved by minimizing error differences between model predictions and observed outputs. An examination of such differences provides the basis for adjustment of process parameters, or minor amendment of model structure. Where possible, it is normal to continue to test the model; this is the verification stage and again may lead to minor changes in parameter values, or even to major changes in model structure. Verification includes the use of data not used in the calibration phase, and may involve new field sites; thus the 'mobility' of the model is also evaluated. Figure 7.1 illustrates the development of a model to the calibration stage, and by implication includes verification also.

Having produced a 'satisfactory' model, it is then usual to use it! Model use may fall into three areas: as a didactic tool (to illustrate how the system works – in part, or as an integrated whole); as a research procedure (to investigate and/or experiment with the system); or as a forecasting instrument (for situations where actual predictions are required).

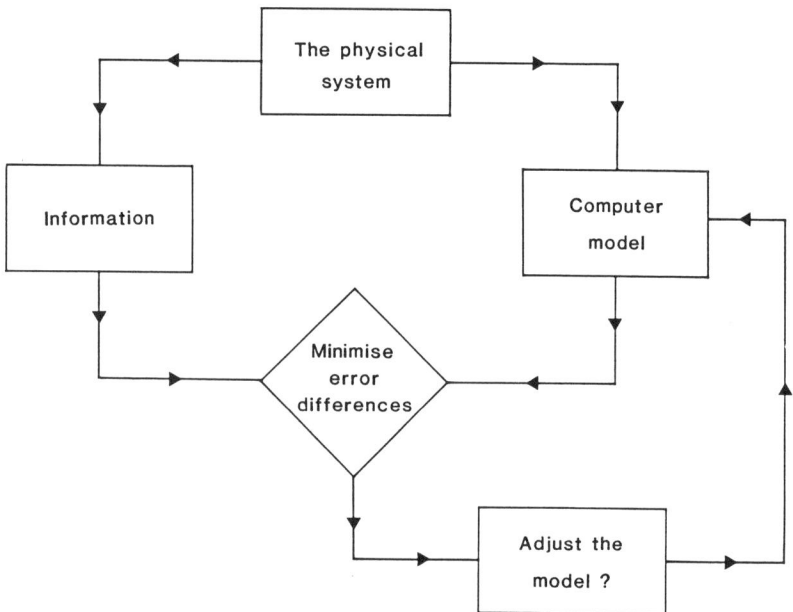

Figure 7.1   Stages in the development of a computer simulation model

It is clear from what has been stated above that the type and quality of data available to the modeller is of the greatest importance. If certain data are unavailable, this may determine which type of model will be developed, or may limit model use to non-forecasting applications. The model itself must be designed using input data and process parameter values which are subject to measurement and sampling errors, and in some cases the process parameters may be invalid in that they do not equate exactly with the field variables on which they are based. Having thus calibrated the model using data which are erroneous to an extent, we then verify the model using recorded data from the physical system which is itself subject to measurement error (if it exists at all). Thus, all stages of model development and use are constrained by the quality of the available data: this point forms a central theme to the detailed discussion of model calibration and verification which follows in this chapter.

7.2 MODEL CALIBRATION

The specific relationship between the general model form and the physical system being studied is gained via the model parameters. It is the accuracy of the parametric values which determines the goodness of fit between the model output and recorded output. Ideally, the model structure should reflect reality as closely as possible, notwithstanding the simplifications of process operation and spatial variability contained within the model. However, the further the model departs from the known physical system, the more tenuous becomes the relationship between model parameters and system characteristics. If the model is to be transferred to an unknown site (e.g. to an ungauged drainage basin), the parameter values must equate with physical properties of the system; as noted above, such parameters are termed 'process' parameters, and may be distinguished from 'optimized' parameters which are usually determined automatically using a computer-based optimization routine. Thus, we can identify three methods of setting the parameter values:

1   We can use field or map measurements. If the model is physically-based, the process parameters have real physical meaning in terms of actual properties measurable in the field, or from maps. In this case the accuracy depends on our choice of model structure and on the quality of data used for calibration. For example, in the synthetic unit hydrograph model (UNIT), the unit hydrograph dimensions are calculated using basin characteristics such as channel gradient and percentage urban area; these are derived from map measurements.

2   We can alter the parameter values so as to optimize model accuracy. This may be done manually by a process of trial and error or we may use automatic computer routines. In either case the optimized parameter value has no real meaning. Fitting a regression line through a scatter of

points is perhaps our most common encounter with the process of optimization. Using STORFLO, the parameter m could be optimized if there is an observed discharge record to compare with the model output; this is 'optimization' and the value of m would have no physical meaning. Alternatively, m could be calculated using stream discharge data; this must be achieved if the model is to be applied to ungauged catchments. In the flood hydrograph model IEM4 (NERC, 1975), there are four parameters which must be optimized: two parameters define the relationship between soil moisture deficit (SMD) and infiltration loss. Whilst it has proved possible to predict their values on the basis of catchment characteristics, the normal practice is to fit a negative exponential regression to observed values of SMD and percentage runoff. The other two parameters define the links between water storage and outflow discharge and have not proved amenable to prediction; they must be optimized. Thus IEM4 remains suitable only for gauged catchments.

3　In some models there will be a mixture of process and optimized parameters. Those established by measurement remain fixed whilst the others may be optimized.

The selection of a criterion of model accuracy is also an important aspect of the model calibration procedure, since it provides the basis for adjustment of parameter values. The various methods used to calculate 'goodness of fit' will be discussed in the next section. The purpose of calibration is to minimize differences between the observed and predicted outputs, and, as noted above, this may be achieved by careful measurement or by optimization procedures. Figure 7.2 demonstrates the principle of optimization using a version of the time–area hydrograph model (TIMAREA). The model was fitted to the Upper Derwent catchment in Derbyshire, England. Stream discharge records were available to compare with the model output. Parameter c, which is a runoff coefficient defining the fraction of rainfall which becomes direct runoff, was optimized by trial and error. Using the optimal value for c of 0.17, the predicted hydrograph mirrors closely the recorded output in terms of both runoff volume and peak discharge (see also Table 7.1). The timing of the discharge peak may be criticized since the predicted peak occurs one hour too early. This illustrates the problem of optimization where the model provides multiple outputs; the modeller must decide which are the most important aspects of the output which must be optimized in preference to less important attributes. In this case, it was decided to minimize differences between discharge values, although fortuitously a value of 0.17 for c also minimized differences between total runoff volume during the simulation period.

The optimization curve on Figure 7.2 takes us on to the question of parameter sensitivity. A sensitive parameter can be defined as one exhibiting a steep error gradient away from the point of minimum error. In other

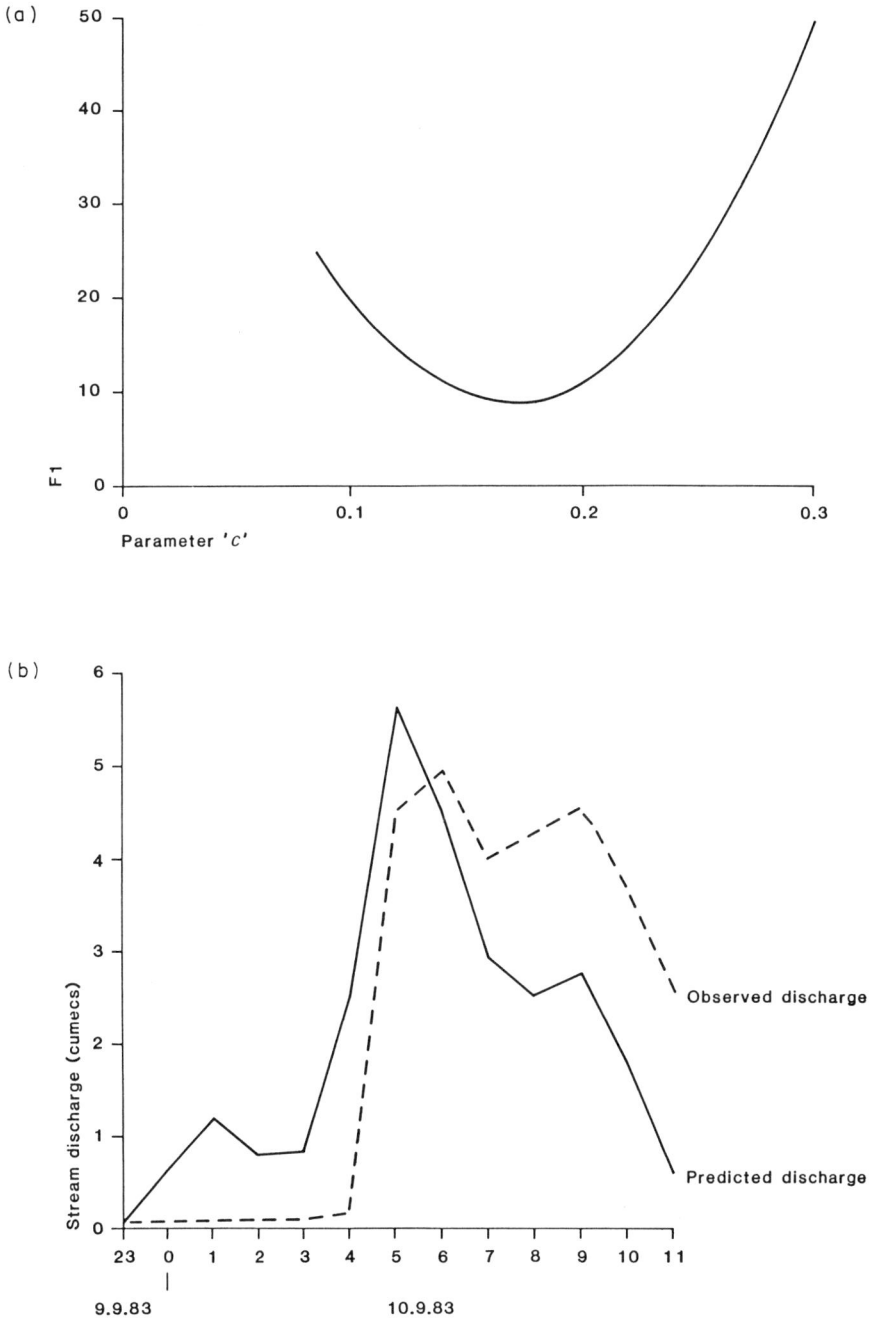

Figure 7.2 Optimizing a parameter value for the time–area hydrograph model TIMAREA (a) a plot of the parameter value c^1 against the error function F_1; (b) the optimized hydrograph compared with the observed hydrograph

Table 7.1 Optimization of the time–area model (TIMA-REA) parameter c for the Upper Derwent catchment, using the storm hydrograph of 9 September 1983

Parameter c	Error statistics	
	F_1	F_2
0.10	20.27	38700
0.125	13.58	25228
0.15	9.84	11761
0.165	9.02	3675
0.17	8.98	990[a]
0.175	9.06	1706
0.18	9.26	4399
0.20	11.24	15174
0.225	16.39	28641
0.25	24.49	42112
0.275	35.56	55584
0.30	49.58	69044

Hydrograph for $c = 0.17$

	Hour	Predicted discharge	Observed discharge
9 Sept	23.00	0.079	0.079
10 Sept	00.00	0.652	0.079
	01.00	1.192	0.080
	02.00	0.805	0.082
	03.00	0.816	0.096
	04.00	2.545	0.179
	05.00	5.666	4.504
	06.00	4.586	4.934
	07.00	2.934	4.044
	08.00	2.533	4.291
	09.00	2.749	4.589
	10.00	1.823	3.698
	11.00	0.605	2.693

[a] Optimal solution.
c = runoff coefficient.
F_1 and F_2 are error functions as defined in the text.

words, a given change in the parameter value will cause a much greater increase in error for a sensitive parameter than for an insensitive one. For a sensitive process parameter, the field-measurement programme must provide a sample of sufficient size to predict accurately the mean parameter value. Conversely, an insensitive parameter is less affected by measurement and sampling errors, and should be given less attention than more sensitive parameters if only a limited field programme is feasible. Sensitivity analysis is normally conducted by assessing the effect on the model output of a fixed percentage change in each model parameter, while holding all other parameter values constant.

An opportunity to conduct a sensitivity analysis is provided by using the program IEM4T (Figure 7.3), which is an addition to IEM4 as presented in Chapter 4 (combined as IEM4TOT). IEM4TOT compares the simulated hydrograph to the hydrograph which would have resulted if all the parameter values remained unchanged. Table 7.2 gives the optimized parameter values for the River Tyne (this is the example given in Section 1.7.3 of the *Flood Studies Report*); also given in this table are the parameter values increased by 10 per cent. Run the model IEM4TOT changing one parameter value at a time. Note the error statistics printed at the end of the program: the largest values correspond to the most sensitive parameters, and vice versa. Having produced your table of error statistics, see if you can explain, in relation to the model structure as described in Chapter 4, why some parameters are more sensitive than others. The error statistics given are a sum of squares difference between observed and predicted hourly discharge values (F), and the difference between the total observed and total predicted runoff volumes for the entire hydrograph (D). These types of error statistics are explained in more detail in the next section.

A sensitivity analysis based on a fixed percentage change for each parameter value may be unrealistic with respect to the range of variation which is observed for each parameter in the field. Thus, an alternative course is to conduct sensitivity analyses using the properties of the frequency distributions for each process parameter, the fixed change being related to the standard deviation of the parameter mean value, for example. Such an approach might indicate new 'sensitivities' in the model structure: some parameters, normally thought to be insensitive, may prove difficult to define accurately in the field and thus provide a real obstruction to model accuracy. Other variables, despite their sensitivity, may be easily measured in the field and thus give us no problem. Once again, the interaction of model structure and data requirements is emphasized. Gardner *et al.* (1980) have provided an example of this approach. Using a marsh hydrology model, they show that sensitivity analysis must focus on both the importance of the parameter within the model mechanism, and on the uncertainty associated with the parameter value. They demonstrate that certain parameters, particularly the upper level of the water table (the level when zero infiltration occurs), can have low sensitivity but contribute large prediction errors as a result of measurement uncertainty. They also demonstrate seasonal variations in parameter sensitivity. Note that if such a sensitivity analysis is carried out before the programme of field measurements, it is possible to use published data in order to estimate the statistical properties of each parameter.

If a model has only one parameter, it is not usually difficult to find the optimal parameter value (except perhaps for the situation where thresholds are involved and the output may be unaffected by the parameter!). This is

162

```
135 DIM P(75): FOR I=0 TO 75: READ P(I):NEXT I
139 REM ===========================
215 TQ=Q(0):TP=P(0):SQ=TQ*TQ:A=TP-TQ:F=A*A:
    REM Initialize Error Sums
220 R=1:A$="Time(Hrs)    Rain(mm) SIM -cumecs- OBS":
    GOSUB 16000:R=2
375 IF K<=75 THEN A=P(K)-Q(K): F=F+A*A:TQ=TQ+Q(K):TP=TP+P(K)
381 NQ=NT+1:: IF NQ>76 THEN NQ=76:
    REM Error function calculations
382 A=0:GOSUB 795:GOSUB 795: GOSUB 795: R=R-3:IF R<0 THEN R=0
383 C=0: A$="Error Statistics":GOSUB 16000:R=R+1
384 A$="F1 = ":GOSUB 16000:C=5:B=7:D=2:A=F:GOSUB 17200
385 C=20:A$="F2 = ":GOSUB 16000:C=25:A=TQ-TP:A=A*A:
    GOSUB 17200
386 R=R+1:C=0: A$="Efficiency =":GOSUB 16000:C=14:B=5:
    A=SQ-TQ*TQ/NQ
387 A=INT((1-F/A)*100):GOSUB 17000
388 REM ===========================
505 FOR X=0 TO NQ-1:Y=P(X):GOSUB 25000: NEXT X
506 A$="Q Obs": GOSUB 23000
610 REM Data for titles
620 DATA Isolated Event Model,for Sensitivity Analysis,"",
    Forecasts runoff hydrograph, from input rainfall sequence,
    "",Enter parameters as follows
630 DATA Basin Area (sq.km),307,150,600
640 DATA % runoff at saturation,54,27,110
650 DATA Soil parameter (mm),220,110,440
660 DATA Residence Time (hr),2,1,4
670 DATA Delay in hours,4.51,2,10
680 DATA Initial Q (cumecs), 2.3,1,5
690 DATA Initial Deficit (mm),68,30,150
700 DATA Daily Evap (mm),1,0,3
710 DATA Total Duration (Hrs),75,20,110
745 READ RR
760 IF RG=2 THEN RA(K)=RR:A$=STR$(RR): GOSUB 16000
790 C=22:B=2:D=2:GOSUB 17200:C=32:A=P(K):GOSUB 17200
795 IF R=22 THEN GOSUB 29000
801 REM ===========================
802 REM "Observed" hydrograph for comparison
803 DATA 2.3,2.26,2.23,2.20,2.15,2.10,2.05,2.00,1.95,1.90,
    1.85,1.81,1.81,1.93
804 DATA 2.05,2.18,2.38,3.31,4.44,5.10,5.51,5.51,5.42,5.26,
    5.30,5.35,5.19,5.11
805 DATA 5.04,4.90,4.87,4.84,4.70,4.57,4.66,5.63,6.87,7.16,
    6.93,6.84,8.10,9.54
806 DATA 11.05,12.82,14.43,16.26,18.67,21.44,23.43,26.58,
    31.03,34.20
807 DATA 35.23,35.16,33.76,31.74,29.89,28.19,26.63,25.19,
    23.86,21.63
808 DATA 21.49,20.43,19.45,18.53,17.67,16.87,16.11,15.41,
    14.75,14.12
809 DATA 13.54,12.98,12.46,11.97
810 REM ===================================
880 DATA 0,0,0,0,0,0,0,0,0,0,0,0,0,0,0,0,0,0,0,0,0
890 DATA 0,0,0,0,0,0,0,0,0,0,0,0,0,0,0,0,0,0,0,0,0
```

Figure 7.3 A version of the flood hydrograph model IEM4 for use in sensitivity analysis (program name: IEM4T)

Table 7.2 Parameter values for use in the sensitivity analysis with program IEM4TOT

Parameter	Optimized value	Value increased by 10 per cent
Area	307	337.7
PERC	0.5381	0.59191
PERI	0.004899	0.0053889
AC	19.4523	21.39753
DELAY	4.5114	4.69254
SMD	67.818	74.5998
Initial discharge	2.2954	2.52494

seen in Figure 72. If two or more parameters are involved, these should ideally be independent of one another, but often interdependence will occur. Figure 7.4 illustrates the range of possibilities which may arise. If parameters are independent, the error function isolines (joining points of equal error) will be essentially concentric (Figure 7.4(a)). If parameters are interdependent, a minimum error 'valley' will occur, within which it may be hard or impossible to define the parameter values unless one is a process parameter and can be physically 'fixed' (Figure 7.4(b) and (c)). In most complex models, some degree of parameter interdependence seems inevitable but renders the predictions no less accurate. It is only if there are sound physical reasons for requiring independent parameters, or if neither of the interdependent parameters can be fixed prior to calibration, that the model structure will have to be altered. Otherwise it is probably best to leave well alone! Figure 7.5 shows the effect that two soil parameters exert on model output, using a version of the hillslope model, LINEAR, described in Chapter 4. Each of the output 'surfaces' was plotted using the results of several hundred model runs, although these take only a few hours to complete, in fact. For 'peak delayed discharge', the output is almost entirely dependent on the hydraulic conductivity of the upper soil layer; the hydraulic conductivity of the lower layer is unimportant until a threshold is reached and basal percolation becomes so rapid that no delayed hydrograph is produced. Note that as the hydraulic conductivity of the upper layer increases, so the peak discharge increases also, since the water movement downslope through the soil is faster. For 'time to peak delayed discharge', lower layer hydraulic conductivity exerts a more important effect, since by increasing percolation, the delayed throughflow response is truncated leading to quicker (but smaller) peaks; of course, the same threshold effect is also present. Though only a schematic model at this stage of development, the results in Figure 7.5 suggest that unless the lower soil layers are thought to have a high hydraulic conductivity, they can be treated as essentially impermeable, allowing us to concentrate on the problem of measuring the upper layer

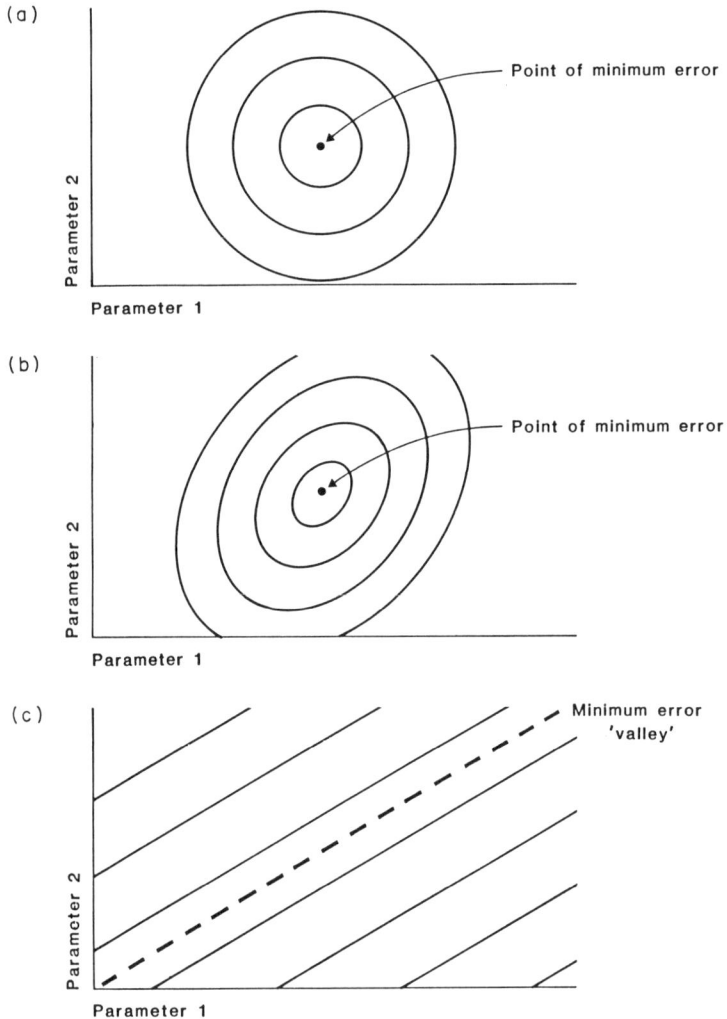

Figure 7.4   Relationships between parameters (a) independence; (b) 'mild' parameter interdependence – a point of minimum error can still be identified; (c) 'severe' parameter interdependence – a minimum error 'valley' exists

hydraulic conductivity in the field. It is perhaps as well to emphasize that this conclusion can be reached before calibration – by running the model using only imaginary, if realistic, parameter values. This shows the value of 'exploratory' simulations, a point which will be developed in Section 7.4.

In conclusion, we should note that sensitivity analysis should be carried out at an early stage in model development, before the field measurement programme is commenced, since this will indicate which of the parameters (both process and optimized) are sensitive, which process parameters may

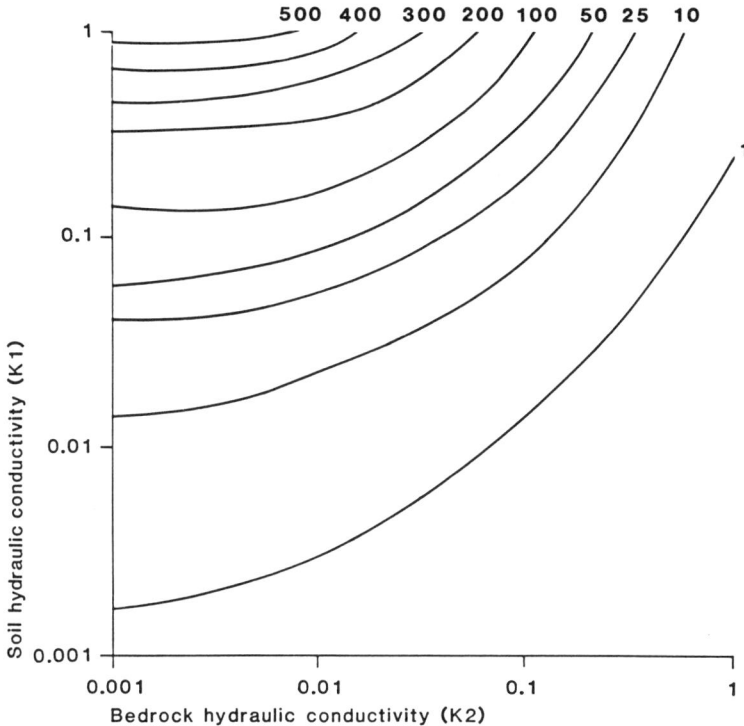

Figure 7.5   The results of a controlled computer simulation experiment, demonstrating the control exerted by soil properties on the peak discharge and timing of a storm hydrograph

require particular attention in the field, and may suggest possible changes in the model structure whilst this is still feasible.

7.3   ASSESSING MODEL ACCURACY

No computer model provides perfect predictions. As we have already seen from Figure 7.1, there is the potential for error and uncertainty both in calibrating the model parameters, and (as we shall see in the next section) in obtaining field data with which to compare the simulated output. It is necessary to develop methods of assessing model accuracy, therefore, in order to reach some objective conclusion as to whether the model is satisfactory or not. This stage of model testing is known as verification.

In general terms, the process of verification involves a comparison of the observed and simulated outputs. Fleming (1975) and Blackie and Eeles (1985) have reviewed the objective functions commonly used to calibrate hydrological models. Most require to minimize the sum of the squares of the differences between the observed and simulated outputs, although there are several ways of achieving this:

$$F1 = s1\,(Q_{\text{obs}i} - Q_{\text{sim}i})^2$$

where

$F1$ = the objective error function
$Q_{\text{obs}i}$ = the observed discharge at time i
$Q_{\text{sim}i}$ = the simulated discharge at time i

$$F2 = (s1Q_{\text{obs}i} - s1Q_{\text{sim}i})^2$$

$$F3 = (Q_{\text{obspk}} - Q_{\text{simpk}})^2$$

where

Q_{obspk} = the observed peak discharge, and
Q_{simpk} = the simulated peak discharge.

$F1$ is the most widely used index, despite the fact that very small timing errors can lead to high $F1$ values. $F2$, a comparison of total storm discharges, and $F3$ a comparison of peak discharges, both avoid this problem of timing but only provide a limited amount of information.

The efficiency of the prediction is given by

$$\text{EFFIC} = \left(\frac{E - F1}{E}\right) \times 100\%$$

where

$$E = s1\,(Q_{\text{obs}i} - Q_{\text{obs}i})^2$$

which compares the prediction with the simplest prediction of all – a constant value $Q_{\text{obs}i}$ at the observed mean discharge. The coefficient of efficiency is useful since it provides a comparison between catchments and between different periods of study, essential if the model is to be genuinely transferable between sites.

Beven *et al.* (1984) have used these goodness of fit measures to test the applicability of the model TOPMODEL (Beven and Kirkby, 1979) to a number of catchments. They record long-term efficiencies of 60 per cent for Crimple Beck and Hodge Beck, rising to 80 per cent for the River Wye where they predicted water balances to within 4.1 per cent of the observed total. Clearly, the assumption in TOPMODEL, that discharge results predominantly from surface runoff generated by a dynamic partial contributing area, would appear to be appropriate in this case. Research in a different area (the Slapton Wood catchment – Butcher, 1985) has shown consistently lower long-term efficiencies, down to 35 per cent, and very much worse water balance predictions. In this instance, because subsurface runoff dominated the storm hydrograph response, TOPMODEL was much less appropriate.

Even for a single variable like stream discharge, there are several aspects of the output that may interest the modeller – total runoff, discharge peaks, timing of peaks, hydrograph shape, and so on. Even for a single output, the modeller must decide which attribute is of most critical interest therefore. Many models produce multiple outputs (e.g. surface flow, subsurface flow, evaporation, transpiration, etc.) so the modeller must select the most important since a single criterion cannot be applied to all these variables. For example, the well-known Stanford Watershed Model (Linsley and Crawford, 1960) recommends the following steps:

comparison of annual runoff volumes
comparison of the seasonal distribution of runoff (daily flows)
comparison of storm hydrograph shape and timing (hourly flows)
comparison of storm hydrograph volume errors

Model verification should be an ongoing process. It is usual to test model accuracy immediately after calibration using a 'split record' test. Available observed data are split into two subsets, one being used for calibration and the other for verification. If the model is stable, verification should result in fully acceptable predictions. Possible reasons for instability at this stage may include the existence of interdependent parameters (see next section), a parameter being inadequately utilized during calibration which suddenly becomes crucial with respect to the independent data (this implies that the calibration data set was not large enough), or some change in the system, either a change in data quality or some real change in the physical system (e.g. a change in albedo following crop growth). If none of these reasons for parameter variation can be substantiated, this may mean that the model itself is at fault: the model can 'fit' only if optimized, but is inadequate as a predictive tool in what is essentially an uncalibrated situation. In such cases where the model is not proven 'robust', it is likely that the model structure needs altering.

7.4 DATA REQUIREMENTS FOR MODELLING

As shown in Figure 7.1, information about the physical system is required for two purposes – to calibrate the model and to test the model performance. In addition, model operation demands a continuing source of available data. Anderson and Burt (1985) considered the quality of information for hydrological forecasting under four main headings – this division is also relevant to simulation models in physical geography in general.

1   The availability of data will determine which class or type of model can be designed and used. In most cases there will be some form of data limitation; this most often seems to apply in the case of meteorological data. For example, the Penman–Monteith model for predicting potential evaporation (see Chapter 3) requires a number of physical measurements, such as net radiation and soil heat flux, which are rarely measured at 'standard' meteorological stations. The absence of such data precludes

the use of this particular model; either an earlier version of the Penman model, which uses 'standard' meteorological observations, must be used, or a simpler class of model, such as Thornthwaite's black box formula, which needs only mean monthly air temperatures and is thus much less data intensive, can be applied. Even where some data exists, it may give insufficient coverage in time or space: such sampling problems are discussed further below.

Whilst the absence of data will severely restrict the ability of the model to provide actual forecasts, it may not prevent valuable modelling from being undertaken, as we have already seen. It is very possible to learn something about the likely range of a parameter's value from examples in the literature. Even if we only guess the parameter values, there is still much that can be done to investigate the theoretical structure of the model and the 'sensitivity' of the parameters.

2   It is important to employ valid measurements. A valid measure is one which measures what it is supposed to measure! The model parameters must equate unambiguously with the variables being quantified in the field. In many cases this presents no difficulty since the 'measurement rules' for quantities such as stream discharge or temperature are clearly defined. Other variables, such as rainfall or soil moisture deficit, are easily measured at a point, but it may be difficult to provide a single value which relates to the wider area being modelled. Finally, some 'conceptual' variables, such as m in the STORFLO model, which represents soil characteristics, may prove difficult or impossible to equate with parameters which can be measured in the field. Either a field variable must be used as a surrogate for the model parameter (e.g. use stream discharge data to quantify m), or the model must be optimized using a computer-based routine, in which case the model parameter loses its physical basis; this loss may or may not be important, as discussed previously.

3   As already noted, the accuracy of field measurements must always be appreciated. The implications of measurement error for model performance have not been much considered, even in hydrological studies where actual forecasts are required (Anderson and Burt, 1985, p. 9). One notable exception to this is the discussion of errors associated with water balance computations for lakes by Ferguson and Znamensky (1981). They show how errors in one element of the balance can endanger the whole analysis, particularly if an important component like evaporation is considered. In the application of the water balance equation there are errors involved in the estimates of all the various components. These errors can be lumped into a net discrepancy term, E, such that

$$s\Sigma\, I + s\Sigma\, O + \mathrm{d}S \pm E = 0$$

where $s\Sigma I$ is the sum of all the input components, $s\Sigma O$ is the sum of all the outputs, and dS is the change in lake storage for the balance period. Some of the errors will cancel out, but a zero value of E cannot give the assurance that all the individual component values are correct; the result may be fortuitous. On the other hand, a large value of E compared to the magnitudes of the other terms does indicate that a serious problem exists in the estimation of at least one of the components. Ferguson and Znamensky show that for lake area A (km^2), the error in the storage change term, E_s (m^3), can be estimated if the error in measuring the water level of the lake, h (cm), is known. If the level can be measured to ± 1 cm, then

$$E_s = \pm 10\ 000A$$

It follows that the relative error B_c of the storage change term compared to the inflow during the balance period, I, is

$$B_c = \frac{E_s}{I}$$

For an average input Q over t days, then

$$B_c = \frac{11.6A}{(Qt)}\ \%$$

If B_c does not exceed 5 per cent, it is comparable to the error involved in measuring the stream discharge inflow itself, and so the error in measuring water level (± 1 cm) would be acceptable. If, however, B_c greatly exceeds 5 per cent, and if the water level measurement cannot be improved, then a longer balance period (t) is needed to reduce B_c. Thus the errors involved in such storage computations are interrelated, and inadequate measurement of just one component can ruin the whole analysis. It is clearly important to minimize errors associated with the largest terms in the balance, and this may determine where most investment is needed in improving field data collection. Such 'sensitivity' in terms of data quality is a neglected aspect of sensitivity analysis in the more general sense: just as a slight change in a parameter value may greatly affect the model output, so measurement error may also affect model performance. Such errors need to be evaluated and, where possible, minimized. Any modelling which relies on grossly inaccurate data cannot be expected to yield accurate forecasts, and might even seriously impair the use of the model for teaching and research purposes. Parameter uncertainty and its implications for hydrological forecasting have been considered by Gardner et al. (1980) and by Beven (1985).

4   Some of the problems of parameter uncertainty are really a question of sampling, i.e. how representative the individual point measurements are. Sampling in both time and space presents problems. In time, sampling error can be reduced by increasing the frequency of observations, and by stratifying the measurement programme to concentrate on periods of

greatest importance. For example, in the Penman–Monteith evaporation model, greater accuracy could result from more frequent observations being made during the day and specifically during certain periods such as when the windspeed exceeds a certain threshold. Of course, to increase the number of observations being made is a potentially costly exercise, a point considered by Armstrong and Whalley (1985) with respect to the design and use of data loggers. Another example, from Walling and Webb (1981), shows how an increased sampling frequency, particularly during periods of storm runoff, produces more reliable estimates of the sediment and solute load of streams; infrequent sampling, even if on a regular basis (say once a week) cannot be expected to yield nearly such trustworthy results. Sampling in space presents no less of a problem. It is not often that physical geographers can meet the demands of statistical theory: to provide a sample of requisite size may be too costly or impossible in the time available. Often logistics preclude the operation of more than one field site and thus point measurements of parameters must suffice. There is now some information available on the field heterogeneity of model parameters, and techniques such as geostatistics may assist in the task of model calibration (see, for example, Trudgill, 1983; Beven, 1985). However, as yet, few studies have specifically considered the implications of field variability with respect to the uncertainty of model predictions, even in hydrology (Anderson and Burt, 1985, p. 9). In the example given below, using the time–area hydrograph model, it will be shown how model accuracy is affected by the availability of a network of raingauges, rather than having to rely on just a single gauge.

We have already noted the interaction between model structure and data quality and the difficulties of optimization which this may present. The following example will hopefully bring some of these points together. Earlier in this section we used the example of the time–area hydrograph model to demonstrate the principle of optimization (Figure 7.2). A version of the model was specifically developed for the Upper Derwent basin in the southern Pennines, England, in order to investigate the effect of spatial variations of rainfall on the storm hydrograph. A network of 17 raingauges provided the input data, 15 of these gauges being inside the catchment itself. In addition, rainfall-radar estimates of rainfall over the catchment were available. The calibration described in Figure 7.2 was produced using data from all the gauges in the catchment, since this was expected to most closely approximate the actual rainfall input to the area.

We might expect that, as our knowledge of the actual rainfall distribution becomes more uncertain, so will our prediction of the storm hydrograph. The data in Table 7.3 resulted from comparing simulated hydrographs, using a single raingauge (plus some simulations with no more than three gauges), with the hydrograph simulated when using all the catchment gauges. The

Table 7.3 Error analysis of simulated storm hydrographs for the Upper Derwent catchment, using the time–area method (the 'observed' hydrograph is taken to be that simulated when all available rainfall information is used)

Gauges used in simulation	F_1	F_2
A1	1.59	11811
A2	4.85	2498
A3	4.86	7052
B4	4.94	13435
B5	3.39	16797
B6	4.86	7052
C7	0.348	4917
C8	1.43	4960
C9	3.07	6206
D10	6.29	4863
D11	3.07	6206
E12	18.40	23922
E13	2.59	1043
E14	2.75	16250
E15	3.43	5295
B5, A3, El4	3.56	12830
Radar-rainfall	54.73	70671
Snake Pass gauge	14.99	12848

Gauge letters indicate altitude: A – lowest; E – highest
F_1 and F_2 are defined in the text.

results define the additional error which results because of the uncertainty which arises when our knowledge of the input conditions (in this case, rainfall distribution) is limited. The error function F_2 is largely dependent on the total rainfall caught by the gauge(s), but is also influenced by the time-distribution of the rain. Thus gauge E12, which recorded a total of 40.7 mm but with 16 mm in one hour, greatly overestimates the total discharge. Gauge A1, whilst having a total of 36.7 mm, has a more even distribution of rain and the overestimation is less. The error function F_1 is less easy to interpret and seems not to depend simply on total rainfall, location within the basin, or gauge altitude. Since the Upper Derwent is only 17 km^2, the gauges all have very similar hyetographs and so their discharge predictions are quite similar too. The results given in Table 7.3 show that gauge C7, which is quite central to the Upper Derwent basin, provides the best predictions where only a single raingauge is available. However, most of the other single gauges are also satisfactory, and it is only D10 and E12, on the extreme east of the basin, which have large error values. In much larger catchments we might expect much more error to occur as our uncertainty about the rainfall distribution increases, as the number of raingauges is gradually reduced. A series of simulations of this sort showed, in the absence of other information, that a single raingauge in the 'upper centre' of the basin provided the most consistently accurate predictions. Note that the rainfall-radar fails to detect the high rainfall intensities generated over the Pennine hills in this storm; consequently the total rainfall is greatly underestimated and the model predictions are inaccurate. The raingauge at Snake Pass, whilst several kilometres from the catchment, gives quite reasonable predictions. This example of error analysis has not yielded perfect results, but it does illustrate the sort of optimization procedures which can be followed. Such calculations are easily incorporated into your programs by adding just a few extra lines of BASIC. The Upper Derwent modelling exercise suffers from several deficiencies which must be noted, even though the quality of the input data is very high: the catchment is too small to exploit fully spatial variations in storm rainfall; the time–area model is an extremely simple one; the model has been calibrated using only one storm event – a minimum of 20 hydrographs would be used in any forecasting exercise (note that a value of $c = 0.2$ resulted from a larger data set); finally, only certain aspects of the output have been considered in this calibration test, with timing and hydrograph shape not examined. It is clear, though, that calibration is an important part of model development and that both model structure and data quality contribute to the accuracy of the model predictions which result.

7.5 MODEL USE

We can identify three areas for model use:

  for teaching
  for research
  for forecasting

Since the first of these forms the basis for this book, there is hopefully no need to expand specifically on this theme at this stage! The role of models in forecasting is clearly a crucial aspect of applied geography. Unlike most of the models presented in this book, most models are designed as forecasting tools and their use in the classroom has, until recently, been generally a later and fortuitous event. It is self-evident that forecasting models are designed to make successful predictions; thus, calibration and verification are crucial elements in model development. The choice of model type is dictated by a simple principle: the simplest model which will provide acceptable accuracy in solving the defined problem should be adopted. More complex models may be more versatile but a black box model may provide better and cheaper predictions. Thus the development of complex process models, being costly and time-consuming, is likely to be limited to those commercial situations where the benefits outweigh the costs (e.g. the costly development of groundwater resources), or to the research institutes. We have already noted the greater potential for application of deterministic process models. Such models, given their theoretical basis also offer much potential for experimental work, substituting the computer terminal for the fieldsite. We wish to stress this aspect of model use since such theoretical 'exploration' can often lead to significant advances in our understanding of how the physical system operates and how better to model and monitor it.

Konikow and Patten (1985), in discussing process-based modelling of groundwater systems, make several comments about the value of simulation modelling as an aid to formulating an improved model of the aquifer. For example, the important influences of temperature differences and aquifer discontinuities on flow in the Madison aquifer (Wyoming, USA) were only recognized and documented as a result of model analysis. Konikow and Patten (1985, p. 239) note that:

> Although it could be argued that the importance of these influences could have been (or should have been) recognized on the basis of hydrogeological principles without the use of a simulation model, the fact is that none of the earlier published studies of this aquifer system indicated that these factors were of major significance. The difference from earlier studies arose from the quantitative hypothesis-testing role of the model.

Differences between observed and predicted output led them to look for reasons for these inconsistencies, and to develop a three-dimensional approach so as to model the system in a better way. Similar experience with another groundwater flow model enabled them to propose an improved measurement system for sampling the movement of a pollution plume through the aquifer.

We have found the use of simulation experiments to be a particularly useful aid both in teaching and research. In many ways, the same aims are involved in both – namely to develop an improved understanding of how the physical system operates. We have already discussed one example of such experimentation: in Figure 7.5 the role of hydraulic conductivity in a two-layer soil is displayed. The results show that the development of large delayed peaks in stream discharge are largely the result of rapid flow in the upper layer. The lower layer is of little importance until it becomes permeable enough to reduce or even prevent the delayed flow response. The results of further investigations using the hillslope runoff model LINEAR are to be found in Burt and Butcher (1985).

Since all of the simulation models given in this book have parameters which can be varied, any of the models could be used to experiment with. UNIT is particularly amenable to such analysis since the equations for predicting unit hydrograph dimensions can be run again and again without progressing through the whole program each time. When asked whether you want 'more hydrograph parameters?', simply type '1' to work through the parameter predictions again. The process parameters are predicted by a multiple regression equation using basin characteristics. In a controlled

174

simulation experiment, we can investigate the role of a given factor, such as 'channel slope', on the overall parameter values. The following procedure illustrates the sort of thing which can be done.

1   Using the basin characteristics listed for the River Rhymney in Table 2.7, hold all values constant except for 'channel slope'. Run the model using the tabulated value of 14.7, and then repeat using the sequence of channel slope values 0.1, 0.5, 1, 5, 10, 20, 50 and 100. You should get the same results as those listed in Table 7.4. The results show that parameters 'time to peak' and 'time base' are particularly sensitive at low gradients, but at high channel slopes the sensitivity is much less – runoff is very rapid in all cases. Increases in channel slope continue to affect 'peak discharge' throughout the range of values used. You may find it helpful to graph your results, using linear or semi-log paper.
2   Now repeat the exercise for the same sequence of channel slope values, but alter the 'fraction of urban area' as well. If you work in 0.1 intervals from 0 to 1, you should eventually end up with three matrices (one for each unit hydrograph parameter); each matrix will be 9 × 11, requiring 99 simulations in all. This may take an hour or two to complete, but you will then be able to plot a 'contour surface', similar to those plotted in Figure 7.5, for each of the parameters. Note that in this case the urban fraction cannot exceed 1, so that a simple threshold exists in this system. In some of the other models the interaction of parameter values and thresholds upon model output may not be so obvious, and to build up, and, more importantly, to interpret the results of experiments like those plotted in Figure 7.5 may take some time to achieve.

The use of a theoretically-based simulation model can greatly improve our understanding of the physical system, highlighting properties of the system which should be predictable but which, given the complexity of the real

Table 7.4   Experimental results using the UNIT model; the effect of channel slope on unit hydrograph parameters

Channel slope	Time to peak	Peak discharge	Time base of hydrograph
0.1	30.36	4.57	76.53
0.5	16.47	8.43	41.51
1	12.66	10.98	31.90
5	6.86	20.24	17.31
10	5.28	26.35	13.30
14.7	4.56	30.50	11.49
20	4.06	34.28	10.22
50	2.86	48.56	7.21
100	2.19	63.20	5.54

world, only become apparent using such a model. Whilst we may be able to write theoretical statements to describe the operation of the system, once we have more than two parameters to consider, we cannot mentally imagine their combined effect on model output (i.e. we can do no more than conceptualize surfaces like those in Figure 7.5). Nor can field research necessarily be much help, since extensive, controlled experiments in the field are prohibitively expensive, and may be rendered impossible by uncontrollable climatic variations. Even using a simple simulation model, we can systematically improve our understanding of complex environmental systems in a way which has not been previously possible.

7.6 CONCLUSION

There has been a proliferation of models over the last 15 years and a commensurate growth of literature in which such models are recommended, partially described, but rarely evaluated. Moreover, most models tend to be site-specific, solely developed to suit the response of the prototype system. Such a parochial attitude is often justified on the grounds of efficiency, since, after all, predictions are required for the specific system being studied. However, little work is usually done to assess the potential mobility of models to new sites. The claim that a given model solves all previously intractable problems can usually be interpreted as the model providing acceptable accuracy for the study site and unknown (untested) accuracy on other systems. The trouble with testing established models at new locations is that such work is essentially confirmatory, or repetitious, rather than exploratory and developmental. Whilst there may be no incentive to apply a given model to new sites, in the long run much greater modelling efficiency will result from the development of more widely applicable models. In hydrology, some progress has already been made in this direction with the production of physically based process models (e.g. Topmodel, SHE, VSAS2), although as Dooge (1981) notes, hydrology is still guilty of producing mostly site-specific models. Where successfully developed, physically based models have the potential to be applied to new sites, can deal with changes in site characteristics and provide reliable forecasts beyond the range of observations used in calibration.

7.7 REFERENCES

Anderson, M. G. and Burt, T. P. (eds) (1985) Modelling strategies. *Hydrological Forecasting*, Wiley, Chichester, 1–13.
Armstrong, A. D. and Whalley, W. B. (1985) An introduction to data logging, *British Geomorphological Research Group Technical Bulletin 34*, Geobooks, Norwich.
Beven, K. J. (1985) Distributed models. In M. G. Anderson and T. P. Burt (eds), *Hydrological Forecasting*, Wiley, Chichester, 405–35.
Beven, K. F. and Kirkby, M. J. (1979) A physically based, variable contributing area model of basin hydrology, *Hydrological Sciences Bulletin*, **24**, 43–69.

Beven, K. J., Kirkby, M. J., Schofield, N. and Tagg, A. F. (1984) Testing a physically-based flood forecasting model (TOPMODEL) for three UK catchments, *Journal of Hydrology* **69**, 119–43.

Blackie, J. R. and Eeles, C. W. O. (1985) Lumped catchment models. In M. G. Anderson and T. P. Burt (eds) *Hydrological Forecasting*, Wiley, Chichester, 311–45.

Burt, T. P. and Butcher, D. P. (1985) On the generation of delayed peaks in stream discharge, *Journal of Hydrology*, **78**, 361–78.

Butcher, D. P. (1985) The field verification of topographic indices for use in hillslope runoff models, Huddersfield Polytechnic, unpublished PhD thesis.

Dooge, J. C. I. (1981) General report on model structure and clarification. In A. J. Ashow, F. Greco and J. Kindler (eds), *Logistics and Benefits of Using Mathematical Models of Hydrologic and Water Resource Systems*. IIASP Proceedings Series, 13, pp. 1–21.

Ferguson, H. L. and Znamensky, V. A. (1981) Methods of computation of the water balance of large lakes and reservoirs, *UNESCO Studies and Reports in Hydrology*, **31**.

Fleming, G. (1975) *Computer Simulation Techniques in Hydrology*. Elsevier, Amsterdam.

Gardner, R. H., Huff, D. D., O'Neill, R. V., Mankin, J. B., Carney, J. and Jones, J. (1980) Application of error analysis to a marsh hydrology model, *Water Resources Research*, **16**(4), 659–64.

Linsley, R. K. and Crawford, N. H. (1960) Computation of a synthetic streamflow record on a digital computer, *Hydrological Sciences Bulletin*, **51**, 526–38.

Konikow, L. F. and Patten, E. P. (1985) Groundwater forecasting. In M. G. Anderson and T. P. Burt (eds), *Hydrological Forecasting*, Wiley, Chichester, 221–70.

Natural Environment Research Council (1975) *Flood Studies Report*, 5 volumes, Institute of Hydrology, Wallingford, England.

Trudgill, S. T. (1983) Soil geography: spatial techniques and geomorphic relationships, *Progress in Physical Geography*, **7**, 345–60.

Walling, D. E. and Webb, B. W. (1981) The reliability of suspended sediment load data. In *Erosion and Sediment Transport Measurement*, IAHS Publication, **133**, 177–94.

CHAPTER 8

Alternative Modelling Styles

8.1 AN APPROPRIATE MODEL

Part I revealed something of the diversity of modelling styles which are commonly used in physical geography. This chapter aims to give some guidelines as to the choice of an appropriate model. It looks at the sort of questions which need to be answered before formulating a model and compares some alternative approaches to the same modelling problem – that of channel flow. Later sections discuss the development of modelling ideas and some of the problems inherent in more complex models.

8.2 CHOICE OF SYSTEM OF INTEREST

Before tackling the formulation and construction of a model (see Chapter 6), careful definition of the system of interest is required. This can be broken down into a series of interrelated questions:

1   What is the purpose of the model?
2   What time and space scales are of interest?
3   Which are the important variables?
4   Which processes are to be modelled?

The success of the model is, to a large extent, dependent on arriving at consistent and perceptive answers to these questions. (A more detailed checklist of questions can be found in Jeffers, 1982.)

The first major consideration in any modelling work is to specify the purpose of the model. For example, if a representative rainfall sequence is required, use may be made of a randomly generated sequence from a hypothetical or observed distribution as was suggested in Chapter 5. This is clearly an appropriate level of model for the question in hand and should produce results sufficiently close to an expected set of rainfall events. It would clearly not be appropriate to model the entire global atmospheric circulation. Indeed, there is no guarantee that this would give any better results! Thus, it is clear that the modelling effort should be related to the purpose of the model to be constructed.

In terms of the boundary of the system of interest, reference must be made to the range of conditions that are to be modelled. For example, a model of evapo-transpiration from different vegetation types is given in Chapter 3. One of its limitations, however, is that it will only predict potential evapo-transpiration rates and, therefore, will only produce satisfactory results in situations where soil moisture does not become limiting. The addition of a soil moisture store is, in the majority of cases, a *necessary* addition to such a system if the model is to have wide application.

A third and key concern is that of scale. This was a point brought out in Chapter 3 where it was stressed that STORFLO was written with specific scales in mind – the processes were not detailed enough to allow a temporal resolution of less than a day and the spatial scale was limited to small catchments by the range of processes modelled: for instance, flood routing down the channel was not included. This sort of scale restriction applies to most models. However, the influence of time and space scales goes further than this. As shown in the next two examples, it helps to define both the variables and processes which are most important.

The discussion of the status of variables in different time scales began with the work of Schumm and Lichty (1965). However, a recent paper by Church and Mark (1980) brings out more fully the length scales and processes appropriate to different time spans. Table 8.1 gives their conclusions as applied to the analysis of fluvial drainage systems. Time periods are designated along the top of the table and a list of forces is given on the left. Lower-case letters indicate the generalized processes involved, while capitals indicate the corresponding length scale. It is clear from the table that certain effects, such as tectonic and eustatic changes, operate over long time scales and for many systems may be treated as constants (they are frozen). Other factors, such as viscosity, are only important for short-term process details so that for a longer period of interest they may be treated as a single mean value (they are relaxed). This, then, helps to define the forces relevant to a particular process. For example, if one is interested in sediment yield, then the forces of viscosity, body and shear gravity are not likely to be important, tectonic–eustatic effects are frozen, and a model of sediment yield should be based on the other dynamic factors – head force and diffusion – as relevant to basin area and a time scale of years.

Table 8.2 (modified from Church and Jones, 1982) provides a second example in which interest is focused on bedforms in both sand and gravel-bed rivers. These are classified according to scale (column 1), and typical wavelengths are given in column 2. Associated time scales are given in column 3. These are related to event time (t_E) – the time taken for a floodwave to pass through a reach (in absolute terms, this itself will be dependent on the scale of the river concerned). Bedforms are also associated with a certain system scale, i.e. a sediment or flow variable which helps to define the bedform. For example, the system scale for macroforms is flow width and one of the typical features is a pool and riffle sequence in which

Table 8.1 Appropriate scales for the analysis of fluvial drainage systems (based on Church and Mark, 1980)[a]

Seconds (time scale)	10^0 Sec	10^2 Min	10^4 Hour	10^6 Synoptic period	10^8 Year	10^{10} Century	10^{12} Millenium	10^{14} Million yr
Dynamic factor:								
Viscosity	subsurface flow PORE SIZE		overland flow OVERLAND FLOW LENGTH		*RELAXED* ────────────────→			
Gravity Body force	wave motions BOUNDARY LAYER		seiche CHANNEL DIAMETER		*RELAXED* ────────────────→			
Head force	discharge CHANNEL DEPTH		event hydrograph LINK LENGTH	hydrograph BASIN DIAMETER		runoff pattern HOMOGENEOUS REGION	*RELAXED* ───→	
Shear force	discharge CHANNEL DEPTH		hydrograph LINK LENGTH		*RELAXED* ────────────────→			
Diffusion	──────── *FROZEN* ───────→			sediment transport BASIN DIAMETER	sediment yield BASIN AREA	denudation HOMOGENEOUS REGION		
Tectonic– eustatic effects				*FROZEN*	neotectonics HOMOGENEOUS REGION		epierogeny MAJOR STRUCTURAL UNIT	

[a]Lower case indicates the generalized process; capitals indicate the corresponding length scale.

the wavelength of the sequence is 5–7 channel widths (Richards, 1982). This implies that if a model of pools and riffles is to be developed, it is important that this relation with flow width should be preserved. The same holds for each of the other system scales with respect to the particular forms in question.

The second half of Table 8.2 reveals the different features associated with each of the scales for sand and gravel-bed rivers respectively. Some attempt to define the relevant processes has also been made based on Leeder (1982) and Naden and Brayshaw (1987). It is clear, especially at smaller scales, that each bedform is related to a specific flow process and that gravel bedforms are modelled in a very different way to sand bedforms simply because of the different influence of large grains on the flow dynamics. Size of sediment, then, is often as important a determinant of process as scale of bedform and must be borne in mind when attempting to define the system to be modelled.

These, then, are just some of the decisions to be taken in defining one's system of interest before work on formulating and constructing the model can begin. Having made these decisions, however, a range of models may still be appropriate.

8.3 ALTERNATIVE TYPES OF MODEL

One of the most common problems associated with channel flow in both the geomorphological and engineering fields is the prediction of flow velocity and depth for a given discharge and channel width under the assumption of a fixed bed. Several alternative approaches to this problem are currently in use.

The most common approach found in fluvial geomorphology is a simple roughness equation – typically the Manning equation:

$$v = \frac{R^{2/3} \, S^{1/2}}{n} \tag{8.1}$$

where

$\quad v \quad$ is stream velocity (m/s)
$\quad R \quad$ is hydraulic radius (m)
$\quad S \quad$ is water surface slope (tangent)
$\quad n \quad$ is a roughness constant

Coupled with the continuity equation

$$Q = wdv \tag{8.2}$$

where

$\quad w \quad$ is channel width
$\quad d \quad$ is flow depth

and assuming that the hydraulic radius is approximated by the flow depth for wide channels, the roughness equation can be rewritten as

$$d = \left(\frac{Qn}{w \, S^{1/2}}\right)^{0.6} \tag{8.3}$$

This type of model for the prediction of flow depth is entirely empirically based and is typical of the black box approach (Chapter 2). It is a lumped model (Chapter 4) providing, at best, a mean value of flow depth for the river reach concerned.

Even in this form, however, it is capable of some refinement. For example, the roughness constant, n, may be derived in a number of ways in which attention is given in varying degrees to the factors affecting roughness – bed material, channel form and sinuosity, and vegetation. One of the ways of estimating roughness is to use a table of values of the sort given by Gardiner and Dackombe (1983) and reproduced here in Table 8.3 in which roughness effects are accumulated in the form

$$n = (n_0 + n_1 + n_2 + n_3 + n_4) \, m_5 \qquad (8.4)$$

where

n_0	is the value for a straight uniform channel
n_1	is the effect of surface irregularities
n_2	is the effect of channel form
n_3	is the effect of obstructions
n_4	is the effect of vegetation
m_5	is channel sinuosity

Other examples use the Strickler equation to predict grain roughness:

$$n = 0.41 \, D_{50}^{1/6} \qquad (8.5)$$

or

$$n = 0.038 \, D_{90}^{1/6} \qquad (8.6)$$

where

D_{50}	is the median grain size (m)
D_{90}	is the grain size than which 90 per cent of grains are finer (m)

or the Limerinos equation which incorporates the effect of relative roughness or the possible drowning out of roughness elements:

$$n = \frac{0.113 \, d^{1/6}}{1.16 + 2.0 \log(d/D_{84})} \qquad (8.7)$$

where D_{84} is the grain size than which 84 per cent of grains are finer (m).

The goodness of fit of these relations has been tested by Bray (1979) for a set of Albertan rivers with straight uniform reaches. The percentage deviations from measured values are given in Table 8.4. The poor fit of these models perhaps indicates the problems associated with defining a roughness constant, the effect of missing variables or the need for a more sound theoretical base (Bathurst, 1982).

One of the effects missing from the Manning equation is the influence of upstream and downstream flows. This is particularly obvious in the case of backwater curves extending upstream from a weir or other control point. Two of the alternative approaches to predicting flow depths take this into

Table 8.2   Hierarchical bedform classification (partly based on Church and Jones, 1982)

Class	Wavelength (metres)	Time scale	System scale	Sand features	Processes	Gravel features	Processes
Microforms	$10^{-2}-10^0$	$\ll t_E$	D	ripples lineations	flow separation (smooth bed)	gravel clusters (roughness elements)	grain–grain hindrance grain–fluid drag
Mesoforms (1)	10^0-10^2	t_E	d	dunes	macro-turbulence, e.g. bursts + sweeps (rough bed)	step-pool systems transverse ribs	variation between subcritical and super-critical flow with or without bed-armouring.
				antidunes	supercritical flow		
Mesoforms (2)	10^1-10^3	$\geqslant t_E$	d	sand waves	grain dispersion bedload transport	gravel sheets	bedload transport
Macroforms	10^1-10^3	$\geqslant t_E$	w	channel bars pools and riffles	gross flow pattern available sediment	channel bars pools and riffles	gross flow pattern available sediment
Megaforms	$>10^3$	regime time	$\geqslant \lambda$	bar assemblages	sediment waves generated by non-fluvial processes	bar assemblages	sediment waves generated by non-fluvial processes

t_E is event time; D is grain size; d is flow depth; w is flow width; λ is meander wavelength

Table 8.3 Values of Manning's n (based on Gardiner and Dackombe, 1983)

Material (n_0)		Surface irregularity (n_1)	
Earth	0.020	Smooth	0.000
Rock	0.025	Minor slumping	0.005
Fine gravel	0.024	Moderate slumping	0.010
Coarse gravel	0.028	Badly slumped or Irregular rock surfaces	0.020

Variation of cross-section (n_2)		Obstructions (n_3) (debris, roots, boulders)	
Gradual	0.000	Negligible	0.000
Alternating occasionally	0.005	Minor	0.010–0.015
Alternating frequently	0.010	Appreciable	0.020–0.030
	0.015	Severe	0.040–0.060

Vegetation (n_4)		Meandering (m_5)	
None	0.000	Sinuosity <1.2	1.00
Low	0.005–0.010	Sinuosity 1.2–1.5	1.15
Medium	0.010–0.025	Sinuosity >1.5	1.30
High	0.025–0.050		
Very high	0.050–0.100		

184

Table 8.4 Comparison of methods for computing average velocity (figures from Bray, 1979)

Equation		Percentage deviation			
		Mean	Standard deviation	Maximum	Minimum
Cowan	(8.4)	−3.3	29.6	83.2	−50.0
Strickler	(8.5)	44.9	43.7	181.9	−18.6
Strickler	(8.6)	37.5	40.9	159.6	−23.1
Limerinos	(8.7)	2.5	28.8	74.4	−41.8

Percentage deviations given by $100(v_c-v_o)/v_o$ where v_c is computed velocity and v_o is observed velocity.

account. In both examples, the framework used is a one-dimensional distributed model in which the river channel is divided into a number of short reaches with constant bed slope, channel width and roughness characteristics.

The first of these incorporates a stochastic element (Yalin, 1971). Based on the idea that large scale roller eddies produce alternate acceleration and deceleration of the flow, the mean velocity at a cross-section can be modelled by a second-order autoregressive model of the form

$$v_j = R_1 v_{j-1} + R_2 v_{j-2} + e_j \qquad (8.8)$$

where

v_j is velocity at the jth position downstream
R_1 and R_2 are partial regression coefficients
e_j is an error term

Typical values of the coefficients are 1.2 and −0.4 respectively with the error terms being drawn from a normal distribution of mean 0.0 and standard deviation 0.02. The scale of the eddies depends on the size of the channel but have a characteristic spacing of $2\pi w$, where w is channel width. A simple algorithm to generate a sequence of velocities and hence depths is given in the program EDDY (Figure 8.1). The model relies on two initial starting velocities and the other parameters of the equation would be derived from series analysis of observed data. (Methods of series analysis are beyond the scope of this book and interested readers should consult Chatfield (1975) or Richards (1979).)

An alternative way of providing better estimates of flow depth is again to use a one-dimensional distributed model but, instead of adding a stochastic element, to add a process element. The main influence on upstream and downstream flows is that of bed topography. Whereas this may be incorporated in a chosen value of roughness coefficient in the Manning equation, another approach to this problem is a physically-based model developed

```
 10 REM Simple second order autoregressive model to predict
    flow depth
 20 GOSUB 10000:PRINT:PRINT"Yalin's model for water depth":
    PRINT
 30 DIM v(20),d(20):REM downstream velocity and depth
 40 RESTORE:READ r1,r2,xb,sd:REM partial regression
    coefficients; mean and standard deviation of error terms
 50 C=0:R=4:A=5:LL=0.1:UL=100:
    C$="Discharge (0.1-100cumecs) = ":GOSUB 12510:q=A
 60 R=5:A=8:C$="Channel width (0.1-100m) = ":GOSUB 12510:w=A
 70 PRINT:PRINT"Initial velocities (v1 and v2 in m/s) ":C=5:
    R=8:UL=5:A=1.5:C$="Enter v1 (0.1-5) ":GOSUB 12510:v(1)=A:
    R=9:A=1.4:C$="Enter v2 (0.1-5) ":GOSUB 12510:v(2)=A
 80 q1=q/w:d(1)=q1/v(1):d(2)=q1/v(2)
 90
100 FOR j=3 TO 20
110    e=0:GOSUB 180
120    v(j)=r1*v(j-1)+r2*v(j-2)+e
130    d(j)=q1/v(j):NEXT
140 GOSUB 18000:PRINT:
    PRINT"Section     Velocity (m/s)  Depth (m)"
150 R=2:FOR i=1TO20:R=R+1:C=0:B=3:D=3:A=i:GOSUB 17300:C=10:
    A=v(i):GOSUB 17200:C=24:A=d(i):GOSUB 17200:NEXT:PRINT
160 END
170
180 REM Calculation of random error term
190 FOR i=1 TO 48:GOSUB 14000:e=e+A:NEXT i
200 e=(e-24)/2*sd+xb:RETURN
210
220 DATA 1.2,-0.2,0.0,0.02
230
```

Figure 8.1 Simple second-order autoregressive model EDDY to predict flow depth

using Bernoulli's energy equation (Richards, 1978) depicted graphically in Figure 8.2 and given by:

$$S_b\Delta x + d_1 + \frac{v_1^2}{2g} = d_2 + \frac{v_2^2}{2g} + S_f\Delta x \qquad (8.9)$$

where

S_b is average bed slope (tangent)
Δx is distance between adjacent sections (m)
d_1 and d_2 are flow depths (m)
v_1 and v_2 are flow velocities (m/s)
g is acceleration due to gravity (m/s^2)
S_f is energy slope (tangent)

Assuming that flow depth and velocity are known at one cross-section and that the energy slope can be estimated from the Manning equation with a suitable value of the roughness coefficient, then for any sequence of bed elevations, the downstream flow depth is given by:

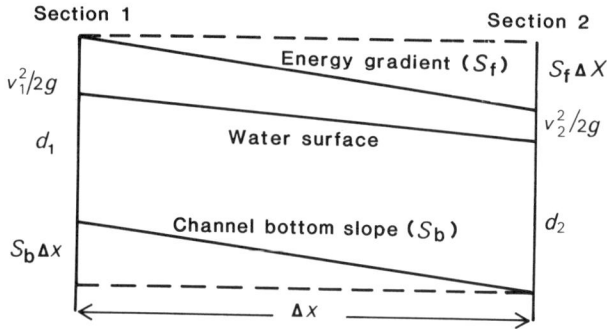

Figure 8.2 Bernoulli's energy equation applied to an incremental reach of open channel (based on Richards, 1978)

$$d_2{}^3 + \left(S_f\Delta x - \frac{q^2}{2gd_1{}^2} - d_1 - S_b\Delta x\right)d_2{}^2 + \frac{q^2}{2g} = 0 \qquad (8.10)$$

where q is the discharge per unit width (cumecs/m). This is a cubic equation in the unknown depth d_2 and cannot be solved analytically. Various numerical methods are available (Scraton (1984) and Section 6.2.1 above); the best of which in this context is Newton's iterative method whereby

$$x_{i+1} = x_i - \frac{f(x_i)}{f'(x_i)} \qquad (8.11)$$

where

x_{i+1} is the estimated solution
x_i is the first guess at the solution
$f(x_i)$ is the function of x_i
$f'(x_i)$ is the derivative of the function of x_i

In this case, the derivative of the function can be written down explicitly as

$$3d_2{}^2 + 2(S_f\Delta x - \frac{q^2}{2gd_1{}^2} - d_1 - S_b\Delta x)d_2 = 0 \qquad (8.12)$$

An algorithm using Newton's iteration technique to solve equation (8.10) allowing 50 iterations for the solution to be found with an accuracy of 0.00001 metres is given in lines 160 to 190 of WATER (Figure 8.3).

Figure 8.4 shows how bed topography might be divided up into uniform reaches, similar to Figure 8.2, each characterized by a length and average bed slope. Negative values of slope are used to indicate a rise in bed elevation in the downstream direction. The program WATER (Figure 8.3) can be used to predict flow depths based on this information (provided in the DATA statements). Flow depths are predicted at 0.1 metre intervals between the endpoints of each segment of the reach. This helps to make the results of such curve fitting more accurate. Readers may like to vary the size of this interval to see the effect on the results for themselves. Graphical results using this technique are given in Figure 8.4.

```
 10 REM Program to calculate water surface profile
 20 REM assumes changes in water depth small enough to allow
    energy slope to be caluated from upstream depth alone.
 30 GOSUB 10000:DIM x(10),sb(10),d(10),y(10),h(10)
 40 PRINT:PRINT "FLOW DEPTH CALCULATIONS":PRINT
 50 C=0:R=4:LL=0.1:UL=100:A=5:
    C$="Discharge (0.1-100cumecs) = ":GOSUB 12510:q=A:R=5:
    C$="Width (0.1-100m) = ":GOSUB 12510:w=A
 60 R=6:UL=10:A=2:C$="Initial depth (0.1-10m) = ":GOSUB 12510:
    d(0)=A:h(0)=0:R=7:UL=0.05:LL=0.01:A=0.03:
    C$="Manning's n (0.01-0.05) = ":GOSUB 12510:n=A
 70 RESTORE:FOR i=0 TO 10:READ x(i),sb(i):NEXT i:
    REM Series of bed distances and slopes
 80 c=q*q/2/9.81
 90 d=d(0)
100 PRINT:PRINT:PRINT"Distance (m) Bed height (m)   Depth (m)":
    R=R+4
110 FOR i=1 TO 10:h(i)=h(i-1)-sb(i)*(x(i)-x(i-1))
120   FOR j=x(i-1)+0.1 TO x(i) STEP 0.1
130      v=q/w/d
140      sf=(v*n/d^0.667)^2:cl=(sf-sb(i))*0.1
150      c2=cl-c/d/d-d
160      FOR k=1 TO 50:f=d^3+c2*d*d+c:fx=3*d*d+2*c2*d
170         dl=d-f/fx:IF ABS(dl-d)<=0.00001 THEN k=50
180         IF d<0 OR k=49 THEN PRINT:
            PRINT"NO SOLUTION POSSIBLE - NEEDS HYDRAULIC JUMP":
            STOP
190         d=dl:NEXT k:NEXT j
200   d(i)=d:C=0:R=R+1:B=3:D=3:A=x(i):GOSUB 17300:C=12:A=h(i):
      GOSUB 17200:C=28:A=d(i):GOSUB 17200:NEXT i:PRINT
210 END
220
230 REM Bed Topography information
240 DATA 0,0
250 DATA 1,0.02
260 DATA 2,-0.01
270 DATA 3,0.05
280 DATA 4,-0.002
290 DATA 5,0.001
300 DATA 6,-0.003
310 DATA 7,0.003
320 DATA 8,-0.02
330 DATA 9,-0.001
340 DATA 10,0.005
350
```

Figure 8.3   Program WATER to fit a water surface profile

In this section three different approaches to the problem of predicting flow velocity and depth have been described. The analysis has progressed from a lumped regression model to a one-dimensional distributed model incorporating either a stochastic element or a simplified process base. In considering the application of these models to the field, it is clear that they illustrate the potential variety of data requirements and parameterization (see Chapter 7) needed in such diverse approaches to the same basic problem. Each has its place depending on the degree of accuracy required, the information available and the focus of attention.

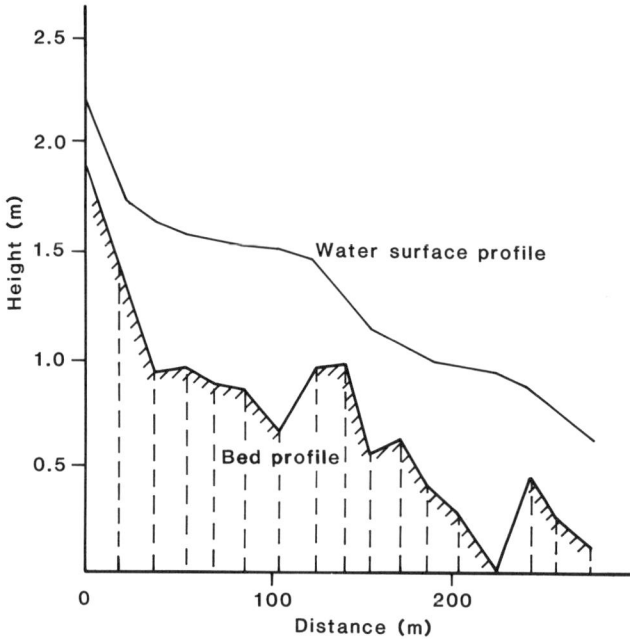

Figure 8.4 Typical water surface profile fitted to a bed profile using the techniques described in WATER (based on Richards, 1978)

8.4 STAGES IN THE DEVELOPMENT OF A MODELLING IDEA

In addition to seeing a range of viable models for a given situation, it is a useful exercise to look at the sort of progression of ideas from simple to potentially more effective models. Here the example used is that of suspended sediment but a similar sequence may be found in many other fields. The most basic model of suspended sediment concentrations is the sediment rating curve:

$$C = a\,Q^b \qquad\qquad (8.13)$$

where

    C is concentration of suspended sediment (mg/l)

    Q is discharge (cumecs)

    a, b are constants from a log-linear regression

This type of model, however, is clearly very rudimentary as seen from the scatter in Figure 8.5(a) – a typical suspended sediment rating curve.

Looking at the time dimension of the problem, however, it is clear that a single-valued relation can only hold if the sediment wave is synchronous with the associated water wave. In most rivers this is not the case and either positive hysteresis (sediment wave in advance of the water wave) or negative hysteresis (sediment wave lagging the water wave) occurs as shown in Figure

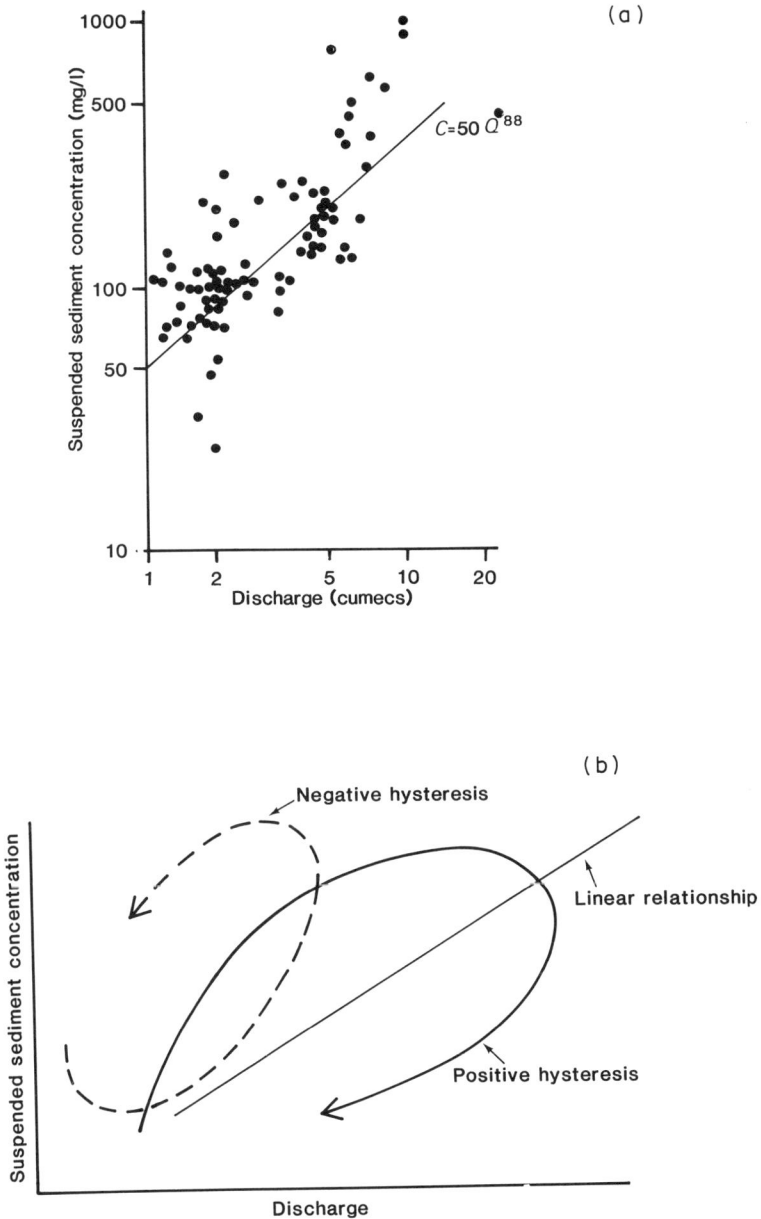

Figure 8.5 (a) Typical suspended sediment rating curve (based on Richards, 1984); (b) types of hysteresis in suspended sediment–streamflow relationships

8.5(b). An example of regression models to describe hysteresis is given by Richards (1984) for a meltwater stream in Norway. Two distinct types of event are recognized: diurnal variation in discharge caused by glacial melting, and storm period variations in discharge caused by rainfall with runoff from the catchment area constituting an important source of sediment. The models derived from events in the 1979 season are

for diurnal variation:

$$\log C = 1.98 - 0.04 \log Q + 1.24 \, \Delta Q \tag{8.14}$$

for storm periods:

$$\log C = 1.51 + 1.20 \log Q + 0.29 \, \Delta Q \tag{8.15}$$

where Q is rate of change of discharge (cumecs/hour). Readers may wish to plot these functions for changing stream discharge to see the hysteretic effects for themselves.

This development of the rating curve idea, by including the time dimension through the influence of change in discharge, is one step towards formulating a more satisfactory model. It simply involves subdividing the observations into more uniformly governed components, e.g. separate rising and falling limbs of the hydrograph; separate events which correspond to different sources of sediment. Further consideration of sediment sources and the processes operating can help to refine the model further.

Positive hysteresis is the more commonly occurring of the two and its main cause is the exhaustion of sediment supply. In other words, it is possible to envisage a sediment store (or series of stores) from which sediment is released in response to a rise in discharge but which, over the course of a hydrograph, becomes depleted and releases less sediment in response to the same level of flow. There are many examples of such stores of sediment – variable contributing areas on the surface of the catchment, available sediment located on previously unwetted banks, or wedges of sediment located behind cobbles on the stream bed itself. Availability and removal of sediment from each of these sources can be modelled by defining which of the sites are accessible to the flow and either a probability of evacuation or a flushing function of the form

$$S = \left(\frac{S_*}{S_0}\right)^n \tag{8.16}$$

where

S	is sediment released to the flow
S_*	is total sediment stored
S_0	is sediment storage capacity
n	is a constant

A flow diagram of the sort of model required to release sediment from cobble-lee storage is given in Figure 8.6. The process part of the model is

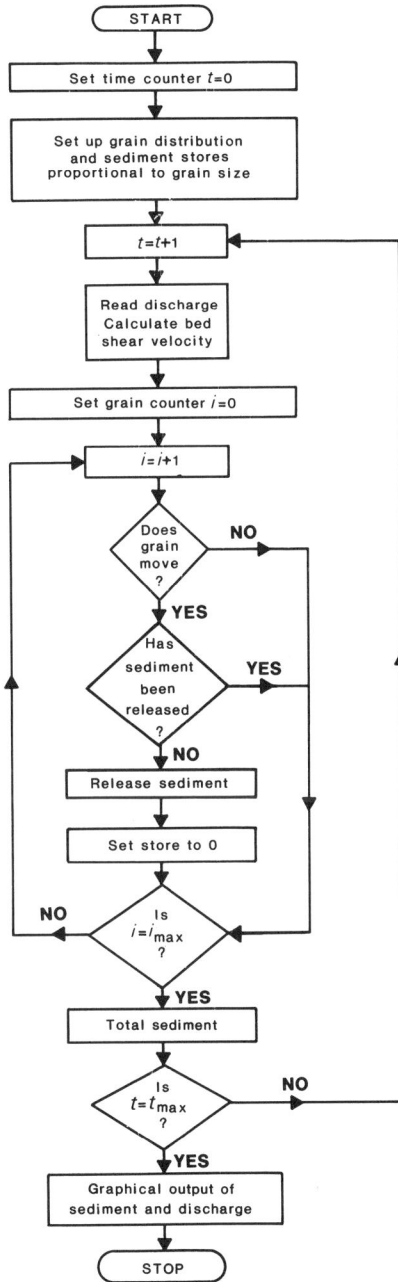

Figure 8.6 Flow diagram for simple cobble-lee release model for suspended sediment omitting recharge of stores (based on McCaig, 1981)

192

a simple gravel movement routine such as that provided by EROSION (Section 3.4). Some results (McCaig, 1981) showing the release of sediment over the hydrograph are given in Figure 8.7.

The sequence of models for suspended sediment discussed above illustrates one of the ways in which modelling ideas develop. In this particular case, the original lumped regression model was first subdivided according to the time dimension and, secondly, subdivided spatially according to sediment sources. In this latter case, there is no fixed physical space scale but the spatial dimension is determined by the geomorphology. Throughout each stage in the modelling, it can be seen that the field problem and data requirements needed to be redefined in order to answer the questions raised by the model.

8.5   MORE COMPLEX MODELS

While most of the examples presented in this book have been simple applications of very basic modelling ideas, the present chapter has begun to discuss how to develop these ideas, and make some progress towards more complicated models. This line of argument relies upon the tacit assumption that more complicated models will be more realistic, more accurate and, therefore, more useful. The validity of this assertion will be examined later. Meanwhile, it may be useful to summarize some of the questions which have to be addressed in the development of these more complex models.

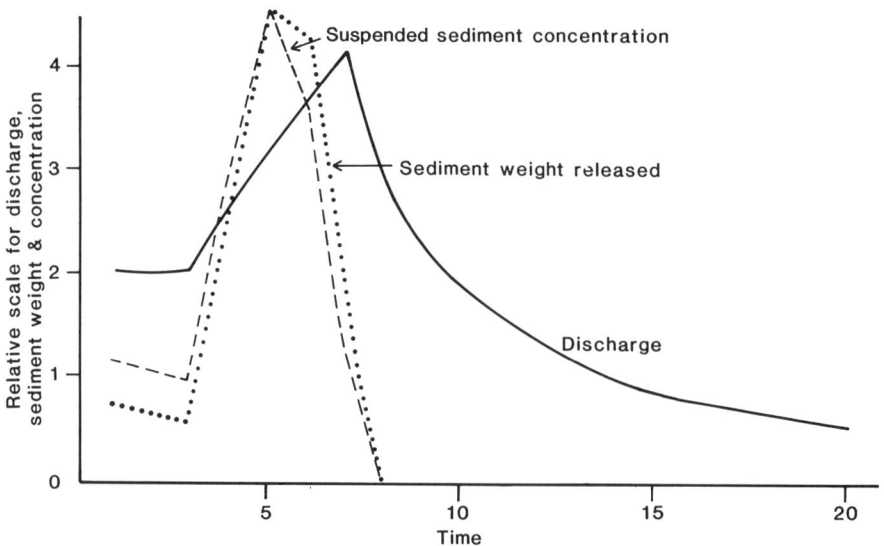

Figure 8.7   Hydrograph and sediment response from cobble-lee sediment release model based on EROSION (based on McCaig, 1981)

Two strategies for increasing complexity have been evident in the examples discussed above. The first of these is the incorporation of more and more *process* detail. This is most clearly demonstrated in the suspended sediment model (Section 8.4) in which a single regression line is replaced by a whole series of process-based models relating to different sources of sediment. A second strategy is the progression from lumped to more and more distributed models (see Chapter 4). This strategy was pursued implicitly in both the water velocity (Section 8.3) and suspended sediment (Section 8.4) models outlined above. In the case of river velocities, modelling progressed from an average velocity for the reach to calculations for successive elements of the reach in the downstream dimension. Conversely, in the suspended sediment predictions, the original rating curve was split in the time dimension to account for rising and falling limbs of the hydrograph, and other such splits could be envisaged, e.g. events governed by rainfall versus those governed by glacier melt (Richards, 1984). However, there are other configurations and the question of dimensions bears further consideration.

In general, the dimensions expressed in a model can range from zero to four including between one and three spatial dimensions and a time dimension plus any number of stochastic degrees of freedom. We have met several models with different dimensions already. The simple regression models such as the suspended sediment rating curve, the summer weather index and universal soil loss equation (Chapter 2), have no time dimension, no explicit spatial application and, apart from the experimental error term in the regression equation, no stochastic degrees of freedom. They are, therefore, zero-dimensional models. The single-store models presented, such as the radiation balance RADBAL (Section 4.2) and the rainfall-runoff model STORFLO (Section 3.2), only operate in the time dimension and are, therefore, one-dimensional models. In model-building we have a choice not only of the number of dimensions to include but also of which dimensions. The examples discussed below illustrate some of the alternatives to the two-dimensional models LINEAR and SLOPEN discussed in Chapter 4 and expressed in terms of a single one-dimensional flow strip through time.

In the model LINEAR, the use of one spatial dimension strictly means that its application is limited to planar slopes. This limitation is reduced by incorporating a factor in the equations to cope with hillslope hollows. The success of this technique really depends upon how important the three-dimensional topography is to the processes in the model. More complex models built on a similar hydrological basis to both LINEAR and STORFLO are more explicit on the role of topography in runoff generation. One of these models is TOPMODEL (Beven *et al.*, 1984) in which the catchment is divided up into subcatchments based upon the river network, with each subcatchment having a relatively homogeneous hydrological response. In a spatial sense, the model is distributed over two dimensions composed of rather irregular areas of uniform topography (represented by the variable $\ln(a/\tan B)$ where a is area drained per unit contour length and B is the slope gradient) and land use. This extension to a three-dimensional model allows both network routing and heterogeneity to be built into the model allowing its application to sizeable catchments, rather than a single hillslope as in LINEAR or very small catchments as in the case of STORFLO. This, therefore, represents a useful expansion of the model.

In the case of slope development models such as SLOPEN, similar in structure to LINEAR, it may be useful to reduce the number of dimensions. For example, the need for a time dimension is quite obvious in cases where interest is focused on amounts of change over different time periods or how long a process takes to reach a particular outcome. However, in many geomorphological models such as those relating to hillslope form, we are interested only in the equilibrium solution to the sediment transport equations. In this case, the time dimension can be dropped as in the analysis of 'characteristic forms' (Carson and Kirkby, 1972).

In expanding slope models into the second spatial dimension as in Ahnert (1976) and Armstrong (1976), a different series of problems can arise. In Chapter 4, the importance of choosing sensible boundary conditions for the models was stressed. In the case of SLOPEN it was assumed that sediment flows were equal and opposite at the divide and that, at the base of the slope, the river carried away all the sediment supplied to it. Other, less simple, boundary conditions could be used – for example constant river downcutting or the absence of a river at the base of the slope (for results with different boundary conditions see Armstrong, 1987). In the case of the catchment topography in three spatial dimensions, care must be taken in explicitly coupling the hillslope and river transport processes.

With increasing complexity, the question as to whether to include stochastic elements in the model must also be answered. In the case of hillslope modelling, micro-topography or slight variations in soil characteristics can conveniently be represented by a stochastic element in the spatial dimension. This was illustrated in Chapter 5 by the generation of different fractal surfaces. In the time dimension, temporal fluctuations on a much shorter time scale than the main processes operating in the model may also be represented by a stochastic routine. One example of this is the addition of a random variable to represent velocity fluctuations in turbulent flow as an extension of models such as EROSION (Section 3.4) to predict gravel movement (Naden, 1987). Quite another reason for wanting some sort of stochastic model is the need to generate distributions of likely variation. In other words, modelling of the real world can never be exact, so a series of model outcomes incorporating some of our uncertainties about the environment can be used to define the range of possible outcomes. One example of this is the application of a form of the random network model (Section 5.5) in Kirkby (1976) where significance levels are quoted for the topological width of generated networks.

These are just some of the additional questions that must be raised in deciding to move towards a more complex model. They basically amount to asking what additional elements are going to yield the most useful information. Having decided on this, the components of more complex models are usually no more involved than the simple examples discussed in this book. They may be derived from experimental or field data. They may be put together from a few assumptions based on physical, chemical or biological laws or just order of magnitude approximations (Harte, 1985). The key to successful models lies in the choice and strength of these components and the skill with which they are put together.

Problems that have to be faced in linking these building blocks may simply come down to computational limitations. So often we want the computer to do a whole series of calculations at once whereas the current generation of machines makes computations in sequence. Getting around this requires careful thought as to how components might be put together and then careful checking as to whether the results are unduly affected by the artificial computational devices used. In other cases, ingenuity is required in careful use of limited computing resources – some hints on saving time and storage space were given in Chapter 6. The only other programming comment is to bear in mind that, with increasing complexity, the need to test the model becomes more and more important, as the results could easily be generated by an undetected programming error, and that the time for debugging the program (Section 6.6) will increase exponentially!

Ultimately, the question is: 'Is it worth it?' The answer to this question may come in a variety of forms but always relating to a cost–benefit analysis of the time, effort and computing resources invested in the study versus the increased accuracy and usefulness of the results or the value of any new insight gained. Three points bear consideration, each illustrated by a different example from the literature.

First, how does the model perform in relation to its data requirements? Here, Beven et al. (1984) present one of the few comparisons of different hydrological models on the basis of their predictive efficiency, as given in Section 7.4, versus the number of parameters that are required to run the models (Figure 8.8). The TOPMODEL relative of LINEAR and STORFLO, represented by the points A, B and C as it was applied to three different catchments, compares quite favourably with other models in that it requires only ten parameters and yields efficiencies of 97.2 per cent, 64.4 per cent and 98.3 per cent in the three cases cited.

A second point in relation to this is the importance of careful sensitivity analysis of more complex models. This sort of analysis was introduced in Chapter 7 but, as the number of model parameters increases, it becomes more and more useful to know how much each parameter contributes to the variance of the results. If all parameters were equally important, the error terms of the model would increase by $\sqrt{m}$, where m is the number of parameters (Miller et al., 1976). Fortunately, this is usually not the case, as illustrated in two well-documented cases – that of mosquito population dynamics, in which 1 parameter out of 29 accounted for 97 per cent of the

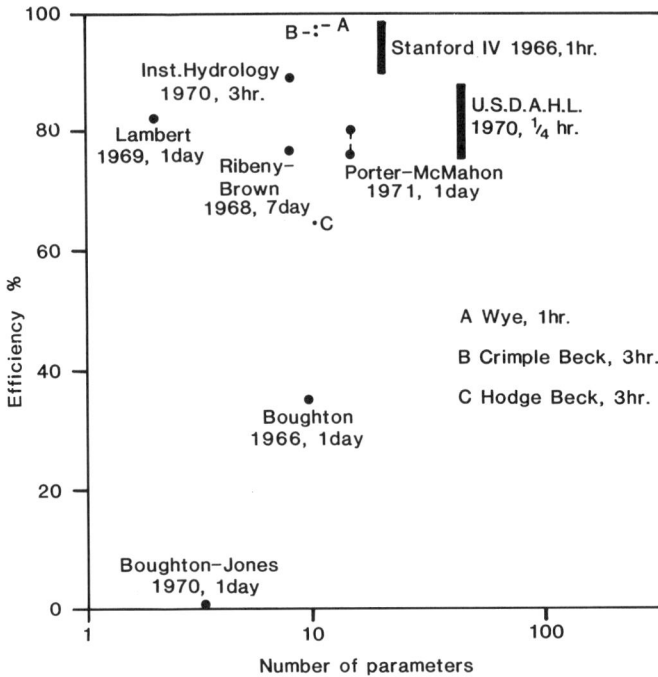

Figure 8.8   Comparison of the efficiency of different hydrological models (based on Beven *et al.*, 1984)

error (Miller, 1974); and that of the distribution and transport of persistent chemicals in the freshwater ecosystem of the Ottawa River, in which 4 out of 27 parameters accounted for 98 per cent of the variance in the results (Miller *et al.*, 1976). Thus, it is apparent that checking the sensitivity of a complex model to its different parameters is an important exercise which can help to simplify data collection and ensure that effort is concentrated on the more critical parameters.

The final point to be considered is a critical assessment of the model itself and, if possible, its simplification. Complex models, as we have said, are made up of a series of blocks derived in different ways and presumably having different degrees of reliability. Just as the sensitivity of the model to its parameters was tested, it is also useful to look at whether there are any of these building blocks which are not important and, therefore, may be omitted from the model without materially affecting its results. Other building blocks may be so unreliable that the next step must be to focus on these parts of the model in order to improve their performance. An example of an assessment of the linkages in a complex model is presented in Naden (in press) for the case of sediment transport in gravel-bed rivers. It is summarized here in Figure 8.9. In short, the final message with all complex models is to refine and, if possibly, to simplify!

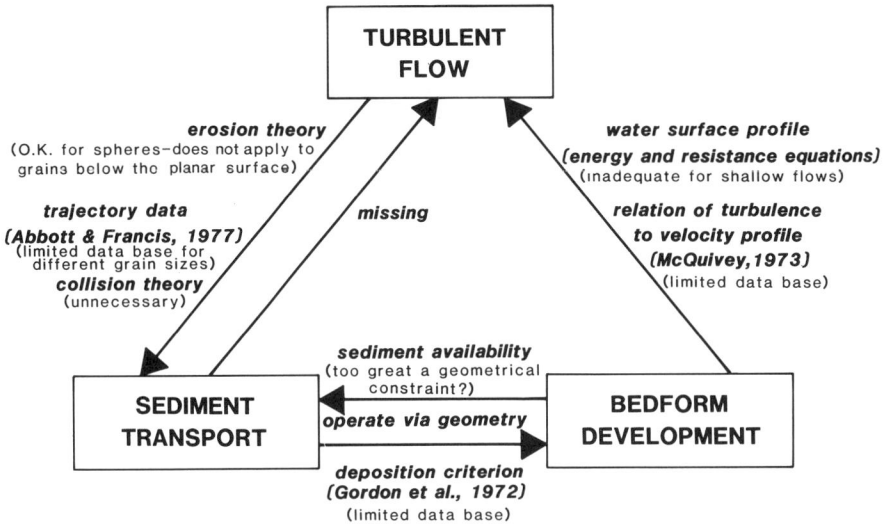

Figure 8.9 Assessment of the links in a sediment transport model for gravel-bed rivers (see Naden, in press)

8.6 REFERENCES

Ahnert, F. (1976) Brief description of a comprehensive three-dimensional process-response model of landform development, *Zeitschrift fur Geomorphologie NF Supplement*, **25**, 29–49.

Armstrong, A. C. (1976) A three-dimensional simulation of slope forms, *Zeitschrift fur Geomorphologie NF Supplement*, **25**, 20–8.

Armstrong, A. C. (1987) Slopes, boundary conditions, and the development of convexo-concave forms – some numerical experiments, *Earth Surface Processes and Landforms*, **12**(1), 17–30.

Bathurst, J. C. (1982) Theoretical aspects of flow resistance. In R. D. Hey, J. C. Bathurst and C. E. Thorne (eds.) *Gravel Bed Rivers*, Wiley, Chichester, 83–108.

Beven, K. J., Kirkby, M. J., Schofield, N. and Tagg, A. F. (1984) Testing a physically-based flood forecasting model (TOPMODEL) for three UK catchments, *Journal of Hydrology*, **69**, 119–43.

Bray, D. I. (1979) Estimating average velocity in gravel-bed rivers, *Proceedings of the American Society of Civil Engineers, Journal of Hydraulics Division*, **105**, 1103–22.

Carson, and Kirkby, M. J. (1972) Hillslope form and process, Cambridge University Press, Cambridge.

Chatfield, C. (1975) *The analysis of time series: theory and practice*, Chapman and Hall, London.

Church, M. and Jones, D. (1982) Channel bars in gravel-bed rivers. In R. D. Hey, J. C. Bathurst and C. E. Thorne (eds.) *Gravel Bed Rivers*, Wiley, Chichester, 291–338.

Church, M. and Mark, D. M. (1980) On size and scale in geomorphology, *Progress in Physical Geography*, **4**, 342–91.

Gardiner, V. and Dackombe, R. (1983) *Geomorphological Field Manual*, George Allen and Unwin, London.

Harte, J. (1985) *Consider a Spherical Cow. A Course in Environmental Problem-solving*, William Kaufman Inc., California.

Jeffers, J. N. R. (1982) *Modelling. Outline Studies in Ecology*, Chapman and Hall, London.

Kirkby, M. J. (1976) Tests of the random network model and its application to basin hydrology, *Earth Surface Processes*, **1**, 197–212.

Leeder, M. R. (1982) *Sedimentology, Process and Product*, George Allen and Unwin, London.

McCaig, M. (1981) Modelling storm period fluctuations in suspended sediment – an appraisal, *Working Paper*, **311**, School of Geography, University of Leeds.

Miller, D. R. (1974) Sensitivity analysis and validation of simulation models, *Journal of Theoretical Biology*, **48**, 345–60.

Miller, D. R., Butler, G. and Branall, L. (1976) Validation of ecological system models, *Journal of Environmental Management*, **4**, 383–401.

Naden, P. S. (1987) An erosion criterion for gravel-bed rivers, *Earth Surface Processes and Landforms*, **12**, 83–94.

Naden, P. S. (in press) Modelling gravel-bed topography from sediment transport, *Earth Surface Processes and Landforms*.

Naden, P. S. and Brayshaw, A. C. (1987) Small and medium scale bedforms in gravel-bed rivers. In K. S. Richards (ed.) *Rivers: Environment, Process and Form*, Basil Blackwell, Oxford.

Richards, K. S. (1978) Simulation of flow geometry in a riffle-pool stream, *Earth Surface Processes*, **3**, 345–54.

Richards, K. S. (1979) Stochastic processes in a one-dimensional series: an introduction, *Concepts and Techniques in Modern Geography Papers*, **23**, Geoabstracts, Norwich.

Richards, K. S. (1982) *Rivers. Form and Process in Alluvial Channels*, Methuen, London.

Richards, K. S. (1984) Some observations on suspended sediment dynamics in Storbregrova, Jotunheimen, *Earth Surface Processes and Landforms*, **9**, 101–12.

Schumm, S. A. and Lichty, R. W. (1965) Time, space and causality in geomorphology, *American Journal of Science*, **263**, 110–19.

Scraton, R. E. (1984) *Basic Numerical Methods*, Edward Arnold, London.

Yalin, M. S. (1971) On the formation of dunes and meanders, *Proceedings of the 14th International Congress of the Hydraulic Research Association, Paris*, **3**, C13, 1–8.

Appendix A

Keywords

The following list shows the BASIC keywords used in the program listings in Chapters 1 to 9. Keywords are in capitals with a brief summary of their use, and variants which may be required for different machines. The list should be read in conjunction with the BASIC manual for any relevant machine, and is not exhaustive in scope. The following notation is used below:

'exp' used for an algebraic expression, variable or constant
'str' used for a string variable or constant
'〈 . . . 〉' indicates an optional part of construction.
'ln' used for a numeric line number.
,   Tabulation print spacer of 8–16 characters.
;   Print next item without space. On some machines prints on next line if no room on current line.
+   Plus sign: also combines strings together. Use '&' instead on QL.

ABS(exp)   Absolute value of expression, with positive sign.
AND   Logical comparison giving true or false. False always has value of 0. True may be −1 (BBC, IBM, Commodore 64, Newbrain) or +1 (Spectrum, QL).
ASC(str)   ASCII code (0–255) of first character of string.
ATN(exp)   Inverse tangent in radians.
CHR$(exp)   String character corresponding to ASCII code for expression evaluating to 0–255.
CLEAR   Eliminates all strings, arrays and variables. Use CLR instead on Commodore 64.
CONT   Continue execution after STOP. **CONTINUE** needed on Spectrum and QL.
COS(exp)   Cosine of angle in radians.
DATA   Announces list of numeric or string values, separated by commas. Strings should be put in quotes for greatest generality.
DEF FNZ(X)   Defines function in single line of program: in this case **FNZ** of variable *x*. Syntax varies: Check manual. Usually use function as **FNZ(x)** or **FNZ(X)**.

DIM A$(M,N)⟨,X(I,J)⟩ Announces dimensions of arrays of string or real variables. **DIM** statement should precede use of array and sets all values to " " (string) or 0 (real). For Spectrum and QL, final dimension of string array sets fixed length of each string.

END Terminates execution: obligatory at end of main program. Replace with **STOP** for QL and Spectrum.

EXP(exp) Raises e (2.718. . .) to the power of the expression.

FOR I=exp1 TO exp2 ⟨STEP exp3⟩ Commences repeated loop. Step of 1 if not stated otherwise. Loop terminated with **NEXT I** (q.v.). For the QL, the first line of multi-line **FOR. . .NEXT** loops should only contain the FOR statement.

GOSUB ln Perform subroutine at ln, which must be a line number actually in the program.

GOTO ln Transfer program control to line number ln, which must be in the program.

IF condition THEN command Jump if condition is met to statements following **THEN** on the same program line. (**ELSE** is not generally supported.)

INPUT variable '?' appears as a prompt to key in a value of the variable. Cues and suggested values supported within PROTO.

INT(exp) The lowest integer below expression (there may be rounding errors).

LEFT$(str,exp) Returns a string of the *exp* left-Hand characters of the string. Not supported on Spectrum.

LEN(str) Returns number of characters in the string.

LET variable = exp May precede assignment of variable value. Compulsory on Spectrum, but not generally used in listings here.

LOAD Load program from disc/tape. Commonly needs to be followed by file name in quotes.

MID$(str,exp1⟨,exp2⟩) exp2 characters of string, starting at exp1th. If exp2 omitted then all characters from exp1th. Not supported on Spectrum.

NEW Clears existing program from memory irreversibly. May be reversed immediately by **OLD** on BBC, but not generally.

NEXT I Terminates **FOR** loops. Variable name (I in example) must be included. It is bad practice to exit before completing full range of loop.

NOT(exp) Returns value 'true' (−1 or 1: see note under '**AND**') if logical expression is false: 'false' (0) if logical expression is true.

ON exp GOSUB ln1,ln2. . . According to value of exp, equal to 1,2,. . ., performs subroutines ln1,ln2,. . . . Not supported on Spectrum, but **GOSUB** exp can be used to give an expression evaluating to ln1, ln2, etc.

ON exp GOTO ln1,ln2. . . According to value of exp, equal to 1,2,. . ., transfers program control to ln1,ln2,. . . . As for **ON. . .GOSUB** above for Spectrum.

OR Logical comparison, returning false (0) or true (−1 or 1: see note under **AND**).

PRINT str or exp, or ; str or exp. . . Prints list of expressions, constants or variables, separated by tabulation (for ',') or without spaces (for ';'). Goes to new line at end of list if not terminated by ',' or ';'. More complex formatting supported through PROTO.

READ Reads next item from data list into named string or real variable name. Types (numerical or string) must match.

REM Remainder of line is treated as a comment, and not executed.

RESTORE Begins **READ** of **DATA** at first **DATA** statement in program. **RESTORE** ln not generally supported.

RETURN Final statement of subroutine to return control to main program.

RIGHT$(str,exp) Returns right-hand exp characters of string. Not supported on Spectrum.

RUN Clears dimensions and variables, and begins program execution with first statement. Variables may be zeroed, but most generally, it should be assumed that they are undefined. To clear variables for QL, begin program with CLEAR (q.v.)

SAVE Followed by file name, usually in quotes. Records program on disc/tape.

SGN(exp) Returns -1 if exp<0; 0 if exp$=0$; and $+1$ if exp>0.

SIN(exp) Sine of expression in radians.

SQR(exp) Positive square root of exp (if positive). Needs to be changed to 'SQRT' for QL.

STEP Optional increment in **FOR. . .NEXT** loops (q.v.)

STOP Temporarily stops program execution. May be resumed with **CONT** (q.v.)

STR$(exp) Converts the numeric value of the expression to the corresponding string.

TAB(n) Prints next text n (numeric) spaces from left of screen. Syntax varies between micros. Better avoided through PROTO.

TAN(exp) Tangent of expression in radians.

THEN Precedes command following **IF**, when condition is met. Command following should be fully explicit, without understood **GOTO**, etc. Use of **THEN** is obligatory.

TO Sets upper limit of variable in **FOR. . .NEXT** loops.

VAL(str) If possible converts the string to a numeric value. For some micros returns a value zero for a non-numeric string, but for others gives an error.

Notes

1 All keywords should be written out in full, and '?' is not an acceptable abbreviation for 'PRINT'. Keywords should appear in CAPITALS.

2 Use line numbers of up to 9000, and 31000 to 32767 for non-executable statements if desired. Other line numbers are reserved for PROTO.

3   Multiple statement lines are allowed, of up to 128 characters.

4   Use one- or two-letter names for real variables: one letter followed by
     '$' for string variables, except I$, J$, K$. Real variable names with letter
     followed by digit, e.g. F7, J2 are reserved for internal use by PROTO,
     as are the string variables I$, J$, K$. Variables for communicating
     parameters to and from PROTO conform to main program naming
     rules.

Appendix B

PROTO Subroutines

B.1 SUMMARY OF SUBROUTINE FUNCTIONS

10000   Initializes machine and sets up text screen. Sets PROTO constants to default values.

11000   Inputs A$, cued by C$ and prompted by current value of A$ at column C, row R.

12000   Inputs A, cued by C$ and prompted by current value of A at column C, row R.

12500   As for 12000, but A is checked to lie between lower and upper inclusive limits LL and UL.

13000   Writes in a function X(, Y(, or Z(as string A$, to be evaluated later in program as FNX(, etc.

14000   A = random value, uniform within range 0 to 1.

14100   A = natural logarithm of A.

14200   A = logarithm to base 10 (common logarithm) of A.

14500   Opens disc/tape file F$ to write to.

14550   Prints directory of current disc to screen.

14600   Opens disc/tape file F$ to read from.

14700   Closes disc/tape file after read or write completed.

14800   Writes numeric value A to disc/tape file.

14850   Writes string A$ to disc/tape file.

14900   Reads numeric value from disc/tape file as variable A.

14950   Reads string from disc/tape as string variable A$.

15000   Opens 40 column text screen: closes graphics screen.

16000   Prints A$ at row R, column C on text screen.

17000   Prints A at row R, column C in ten-digit free format.

17200   As 17000, but with B digits before decimal: D after.

17300   As 17000, but as B-digit integer.

17400   As 17000, but in scientific format: B digits before decimal; D after, followed by 'E' and two-digit exponent.

17500   As 17000, but in B-digit free format.

18000   Clears text screen.

19000   Increments row, R. If at bottom of screen, clears text screen and zeros R on command.

19010 Clears text screen and zeros R on command.
20000 Opens graphics screen with 3–4 line text screen.
21000 Scales graphics screen to ranges NX to XX, NY to XY.
22000 Draws x and y axes, scaled at intervals SX, SY.
23000 Prints A at coordinates X,Y on graphics screen.
24000 Prints A$ on text screen associated with graphics.
25000 Plots a dot at X,Y on graphics screen.
26000 Draws a line from last point to X,Y. Set FG=0 and GOSUB 26000 to start a new line from X,Y.
27000 Plots and fills a rectangle between X and XA, Y and YA.
28000 Clears graphics screen.
29000 Dumps text screen to printer if connected.
29500 Dumps graphics screen to printer if connected.

B.2 DETAILED SPECIFICATION OF SUBROUTINES

10000 Initializes machine and sets up text screen; sets PROTO constants to default values

Input parameters: None

Functions: Checks whether printer attached. Sets flag F7 to 1 for printer use: 2 if not in use.
Opens text screen 40 columns × 24 rows. Sets flag F9 to 1 for text screen.
Sets print format to default (ten-digit free format).
Sets row and column variables R, C to default values of 999.
  With these default values output goes to current cursor position. Resets both R and C to 999 to return to default state.
Sets A$, C$ to empty string, " ". Sets A to zero.
Sets default upper and lower limits for input through SUB 12500 to LL=0: UL=100.
Sets default ranges for graphics screen of NX=NY=0: X=XY=0.
Sets flag A9=0 to indicate that disc/tape file not open.

Output parameters: None, but flags F7, F9, A9 may be used to check status.

11000 Inputs A$, cued by C$ and prompted by current value of A$ at column C, row R

Input parameters: C$ = string to prompt entry, e.g. "Enter required month": A$ = suggested string as prompt, e.g. "March": C,R = column and row to begin printing C$ on screen.

Functions: Prints C$ and then A$ without a space. Returns cursor to beginning of A$ to await acceptance or correction of A$. Editing functions vary with machine. For BBC, accepts original string or modified string, whichever is longer. F1 and F2 keys give cursor left and right functions during editing of A$.

C$ is printed starting from row R, column C (top row and left-hand column are R,C=0). If C=999 and R=999 then C$ printed at current cursor position.

Output parameters: A$ takes new value of input string: C, R and C$ are unchanged.

12000 Inputs A, cued by C$ and prompted by current value of A at column C, row R; exactly the same as 11000 except that a numeric value A is prompted and returned in place of A$ in 11000

Input parameters: C$= prompt string: R,C = row and column to print C$: A = suggested value.

Function: As for 11000. If an invalid number is entered, zero is returned.

Output parameters: A takes new value from input: C,R and C$ are unchanged.

12500 As for 12000, but A is checked to lie between lower and upper inclusive limits LL and UL

Input parameters: As for 12000, but in addition LL = lowest acceptable value: UL = highest acceptable value.

Function: Similar to 12000. Cue takes modified form: for example if C$='Slope length (m)', LL=20 and UL=200, cue appears as 'Enter Slope length(m) (20–200):'. If input value exceeds UL or is less than LL, then input request is repeated at the same screen location as before.

Output parameters: A takes new value within specified range: C,R, C$, LL and UL are unchanged.

13000 Writes in a function X(, Y(, or Z(as string A$, to be evaluated later in program as FNX(, etc.

Input parameter: A$, which should have the form "X(var)=. . .", "Y(var)=. . ." or "Z(var)=. . ." where 'var' is a valid variable name and '. . .' consists of a valid BASIC expression containing the variable name 'var', and other variables or numeric values which have previously assigned values. With the addition of "DEF FN" to the front of A$, the whole string should be a valid function definition in BASIC.

Function: Best shown by an example. If we program:

$$A\$="X(T)=5*T+P": GOSUB\ 13000$$

then if P has been assigned the value of 2, subsequent calls of 'Y=FNX(3)', 'Y=FNX(P+4)' will respectively return Y as 17 and 32. Only the three numeric functions X,Y and Z may be set up in this way, and each may have only a single parameter, 'var'. In many implementations, although not on the BBC, this subroutine requires disc access, with program merging. For the QL, this call should not be made within a subroutine: 'GOSUB 13000' should also be followed invariably by ': MERGE QL$'

Output parameters: None: A$ unchanged.

The next three subroutines are a group of standard functions for which the syntax varies between micros.

14000 A = random value, uniform within range 0 to 1

Input parameters: None.

Output parameter: A = random value between 0 and 1, drawn from a uniform distribution.

14100 A = natural logarithm of A

Input parameter: Positive number, A.

Output parameter: A = natural logarithm of input value.

14200 A = logarithm to base 10 (common logarithm) of A

Input parameter: Positive number, A.

Output parameter: A = logarithm to base 10 of input value.

The following is a group of file functions which vary between machines. Only one such file may be opened at any time. The flag A9 is used for the channel number open, or takes the value zero if a file is not open. F$ is used to identify the correct file on opening for read or write.

14500 Opens disc/tape file F$ to write to

Input parameter: F$ = filename to be opened.

Function: If filename requires machine specific extensions or file device names, these are added. File is opened to write data into, provided that no other file is already open, and that filename does not already exist.

Output parameters: None, but A9 can be accessed as channel number open and/or file status: F$ is unchanged.

14550 Prints directory/catalogue of current disc to screen

Parameters: None.

Function: Prints directory to screen, to allow choice of appropriate file to load, or to create a file avoiding duplication. Directory lengths and formats may vary, so that screen position is generally lost after this call.

14600 Opens disc/tape file F$ to read from

Function: As 14500, except that file is opened to read data from, and that file must already exist. Values or strings are read from the start of the file.

14700 Closes disc/tape file after read or write completed

Input parameters: None.

Function: Closes the previously opened file at the end of input or output sequence.

14800 Writes numeric value A to disc/tape file

Input parameter: Numeric value A.

Function: Writes A to the open file as a string. Takes no action if file not open to write data.

Output parameters: None: A is unchanged.

14850 Writes string A$ to disc/tape file

Function: As 14800 except that string variable A$ is written.

14900 Reads numeric value from disc/tape file as variable A

Input parameters: None.

Function: Reads a string from file, if open to read from, and assigns it to string variable A$.

Output parameter: String A$.

14950 Reads string from disc/tape as string variable A$

Function: As 14900 except that string from file is converted to a numeric value and assigned to A.

15000 Opens 40-column text screen: closes graphics screen

Function: Opens 40×24 text screen: closes graphics screen: sets flag F9=1.

Parameters: None but F9 may be use to check status.

16000 Prints A$ at row R, column C on text screen

Input parameters: A$ = string to be printed: R,C = row and column to print at.

Function: A$ printed at row R, column C. If there is insufficient space on line, text will be moved to left as necessary. If C=999 and R=999 then A$ is printed at current cursor position.

Output parameters: None: A$,C and R unchanged.

These routines are similar to 16000, but are used to output numeric values in various formats. After each use, the format reverts to the default of ten-digit free format.

17000 Prints A at row R, column C in ten-digit free format
17200 As 17000, but with B digits before decimal: D after
17300 As 17000, but as B-digit integer
17400 As 17000, but in scientific format: B digits before decimal: D after, followed by 'E' and two-digit exponent
17500 As 17000, but in B-digit free format

Input parameters: Numeric value A to be printed: B and/or D if required.

Function: The following table shows examples for values printed by the various subroutines. Each column is printed for the same column value, C indicated by the *s in the top row.

Subroutine	*	*	*
SUB 17000	1234.5678	−1234.5678	0.1234
SUB 17200:B=4:D=2	1234.57	−1234.57	0.12
SUB 17200:B=6:D=1	1234.6	−1234.6	0.1
SUB 17300:B=4	1235.	−1235.	0.
SUB 17400:B=4:D=2	1.23E3	−1.23E3	1.23E-1
SUB 17400:B=6:D=1	1.2E3	−1.2E3	1.2E-1
SUB 17500:B=4	1235	−1235	0.123
SUB 17500:B=7	1234.568	−1234.568	0.1234

Output parameters: None: A is preserved and format returns to default of ten-digit free format.

18000 Clears text screen

Parameters: None.

Function: Clears text screen, or text part of graphics screen.

19000 Increments row R. If at bottom of screen, clears text screen and zeros R on command

Input parameter: Row R.

Function: Increments row R. If R near the bottom of the screen, displays message "Press RETURN to continue" on bottom line. On pressing RETURN key, screen is cleared and R is zeroed.

Output parameter: Updated value of R to proceed with printing or data input.

19010 Clears text screen and zeros R on command

Function: Displays message and zeros R as in 19000 irrespective of original value of R.

20000 Opens graphics screen with 3–4 line text screen

Input parameters: None.

Function: Opens graphics screen at best available resolution, usually with at least 256×200 pixels and a monochrome image. Sets flag F9=2 to indicate that screen is opened, but unscaled. Leaves a 3–4 line text screen at top of screen which may be cleared (GOSUB 18000) or written to (GOSUB 24000) separately.

Output parameters: None, but F9 may be accessed to show status.

21000   Scales graphics screen to ranges NX to XX, NY to XY

Input parameters: Minimum NX and maximum XX values of *x* to be shown on graph. Minimum and maximum values of *y*, NY and XY.

Function: Subsequent calls to subroutines 22000, 24000–27000 to plot points, lines and blocks relative to ranges selected. NX and NY default to 0: XX and XY to 1. The status flag F9 is set to 3. If graphics screen is not open, SUB 20000 will be called first. Selects default values of SX, SY for use by SUB 22000.

Output parameters: None. NX, XX, NY and XY unchanged.

22000   Draws *x* and *y* axes, scaled at intervals SX, SY

Input parameters: SX = scale interval for *x*-axis: SY = scale interval for *y*-axis.

Function: Draws axes through the origin, if on screen. Shows scale marks along axes if shown, or along screen edges if axes not shown. The status flag F9 is set to 4. If graphics screen not open and/or not scaled, SUB 20000 and 21000 are called as needed. If SX, SY are not specified, uses default values calculated in SUB 21000.

Output parameters: None, SX,SY unchanged. Status can be checked through F9.

23000   Prints A$ at coordinates X,Y on graphics screen

Input parameters: X,Y are coordinates as scaled on screen: A$ is string to be printed.

Function: String is located with its top left-hand corner at coordinates X,Y. If text cannot be printed to screen, defaults to text in text window.

Output parameters: None. X,Y and A$ unchanged.

24000   Prints A$ on text screen associated with graphics

Input parameter: A$ is string to be printed.

Function: Prints A$ on next line of text screen associated with graphics screen. Should be used in preference to SUB 16000 for printing titles, etc. when graphics screen is open.

Output parameters: None. A$ is unchanged.

25000 Plots a dot at X,Y on graphics screen

Input parameters: Plotting coordinates X,Y.

Function: Plots a dot at (scaled) coordinates X,Y on graphics screen.

Output parameters: None. X,Y unchanged.

26000 Draws a line from last point to X,Y. Set FG=0 and GOSUB 26000 to start a new line from X,Y

Input parameters: FG is status flag, which must be set to zero to begin a new graph: X,Y are coordinates of point to plot line to.

Function: Draws a line to X,Y from previous point called with SUB 26000. If FG=0, begins a new graph starting from X,Y. Uses status flag FG to check for off-screen points, etc. as necessary. Should draw lines to and from edges of the screen where necessary.

Output parameters: None. X,Y unchanged. FG reflects status (=1 if X,Y off-screen: =2 if on-screen).

27000 Plots and fills a rectangle between X and XA, Y and YA

Input parameters: X,XA are left- and right-hand coordinates of rectangle (in either order): Y,YA are top and bottom coordinates.

Function: Draws an upright rectangle. If possible does not hide text or graphs already plotted.

Output parameters: None. X, XA, Y, YA are unchanged.

28000 Clears graphics screen

Funtion: Clears graphics screen but not associated text screen. Does not affect screen scaling, but wipes out axes, etc.

29000 Dumps text screen to printer if connected
29500 Dumps graphics screen to printer if connected

Function: If printer is enabled (i.e. printer flag F7=1) then these routines respectively dump the text screen and the graphics screen with associated text, to an Epson FX80 compatible printer. Code will need to be changed for other types of dot matrix printer. Daisy-wheel printers will not handle graphics dumps.

The implementation of all these subroutines varies a little between machines, as may be seen below and must be adapted to each particular type of printer.

B.3   USE AND LISTING OF PROTO FOR THE BBC MICROCOMPUTER

To load self-contained programs

Type LOAD in capitals, followed by program name within quotes; e.g.

LOAD "GRAPH"

This method applies to all listings in the disc catalogues except for those ending in the letter X, and the program IEM4T. In all cases except the subroutine MATINV the program PROTO needs to be merged with the main program to provide the input and output routines described above.

To merge two programs

There are two methods, with different advantages and limitations, the *EXEC method and the *LOAD method.

*EXEC method: Where the program to be added has been listed in the form of a *SPOOL file, and is listed on the BBC disc for this book with a file name ending in the letter X, then the program is merged by typing *EXEC followed by the file name (ending in X), but not in quotes. The appropriate file for the PROTO program is called PROTEX, so that it may be merged to any program by typing:

*EXEC PROTEX.

This method may also be used to merge STR_AN with CHSTR or CHNET by typing:

*EXEC STRANEX

This may be done either before or after merging with PROTO/PROTEX using either method.

   This method may also be used to merge IEM4T on to IEM4, by typing:

*EXEC IEM4X

The merged program is already on disc as IEM4TOT, which still requires PROTO/PROTEX to be merged on to it.

LOAD method: This method is quicker for a long program, and requires no special listing of the program file; but it only works provided that all the line numbers in the second program are higher than all the line numbers in the first. This condition is always met for the listed programs, so that PROTO may be merged in this way for all the cases given. To merge PROTO on to the normally loaded first program, there are three steps to be typed in turn, each followed by RETURN:

1   PRINT ~TOP-2
2   The response to this request is a string representing an address (in hexadecimal). If, for example it is '2C84', then type:

          *LOAD PROTO 2C84

3   END
This final step is needed to integrate the two programs, after which the merged program can be run or listed in the normal way.

Further details of these two methods are given in the *User Guide* to the BBC micro, together with details of how to produce a suitable file for merging using *EXEC.

Appendix B3: Figure 1. Listing for 'PROTO' for the BBC micro

```
10000 MODE7:F9=1:@%=10:VDU 20:NX=0:XX=1:NY=0:XY=1:C=999:R=999:
      LL=0:UL=100
10015 *KEY 1 |H
10020 *KEY 2 |I
10025 *KEY 0 |M MODE 7|M|NLIST,9000|M
10030 A=0:A$="":C$="":A9=0
10040 PRINT TAB(0,5);"Press P if printer connected":PRINT
      "       X if not connected"
10050 A1$=GET$:F7=INSTR("PX",A1$):ON 1+F7 GOTO 10050,18000,
      18000
11000 A0$=A$:L1=LEN(A0$):GOTO12020
12000 A0$=STR$(A):L1=LEN(A0$)
12010 IF LEFT$(A0$,1)=" " THEN A0$=MID$(A0$,2):L1=L1-1:
      GOTO12010
12015 GOSUB 12020:A=VAL(A$):RETURN
12020 FOR I1=1TO L1:A0$=A0$+CHR$(8):NEXTI1:A0$=C$+A0$
12030 A5$=A0$:GOSUB 15150: PRINTA0$;:X1=POS:Y1=VPOS:L2=0
12040 A0=GET:IF A0<>13 THEN VDU A0:L2=L2+1+2*((A0=8) OR
      (A0=127)):L1=L1+(L1-L2)*(L2>L1):GOTO12040
12050 VDU31,X1,Y1:A$="":FOR L2=1 TO L1:A%=135:A0=USR(&FFF4):
      A0=A0 AND &FFFF
12100 A0=A0 DIV &100:A$=A$+CHR$(A0):VDU 9:NEXT:PRINT:RETURN
12500 C$="Enter "+C$+" ("+STR$(LL)+"-"+STR$(UL)+"): "
12510 GOSUB 12000:IF A<LL OR A>UL THEN GOTO 12510 ELSE RETURN
13000 X1%=ASC(A$)-87: IF X1%<1OR X1%>3   THEN PRINT"Functions
      for X( to Z( only":RETURN
13010 A1$=MID$(A$,INSTR(A$,"=")+1):ON X1% GOTO 13016,13017,
      13018
13016 XFN$=A1$:RETURN
13017 YFN$=A1$:RETURN
13018 ZFN$=A1$:RETURN
14000 A=RND(1):RETURN
14100 A=LN(A):RETURN
14200 A=LOG(A):RETURN
14500 IF A9=0 THEN A9=OPENOUT(F$): RETURN ELSE RETURN
14550 *CAT
14560 RETURN
14600 IF A9=0 THEN A9=OPENIN(F$): RETURN ELSE RETURN
14700 IF A9<>0 THEN CLOSE£ A9:A9=0: RETURN ELSE RETURN
14800 IF A9=0 THEN RETURN ELSE PRINT£ A9,A: RETURN
14850 IF A9=0 THEN RETURN ELSE PRINT£ A9,A$: RETURN
14900 IF A9=0 THEN RETURN ELSE INPUT£ A9,A: RETURN
14950 IF A9=0 THEN RETURN ELSE INPUT£ A9,A$: RETURN
15000 MODE7:F9=1:RETURN
15100 C2=37-@% AND 255:IFA5>=10^(@% AND 255) AND @% DIV &10000
      AND &FF <>1 THEN C2=C2-4
15110 IF C=999 AND R=999 THEN C1=POS:R1=VPOS:GOTO 15130
15120 C1=C+C*(C<0)+(C-C2)*(C>C2):R2=24+21*(F9>1):R1=R+R*(R<0)
      +(R-R2)*(R>R2)
15130 PRINTTAB(C1,R1);:RETURN
15150 C2=INSTR(A5$+CHR$(8),CHR$(8)) -1:C2=40-C2:C2=C2+(C2=40)
      +C2*(C2<0):GOTO15110
15200 A$=MID$(B$,M,1):RETURN
15300 A$=MID$(B$,M,N):RETURN
15400 A$=LEFT$(B$,N):RETURN
15500 A$=RIGHT$(B$,N):RETURN
16000 VDU4:A5$=A$:GOSUB 15150:PRINTA$;:RETURN
17000 A5=A
17010 VDU4:GOSUB 15100:PRINTA5;:@%=10:RETURN
17200 @%=B+D+2+256*(D+512):GOTO17000
17300 @%=B+2+&20000:GOTO 17000
17400 @%=B+D+5+256*(D+257):GOTO 17000
17500 @%=B: A7=ABS(A):A5=10^(INT(LOG(A7)))
```

```
17510 A6=10^(B-1+((B-A5)>1)):A5=SGN(A)*INT(A7/A5*A6+.5)/A6*A5:
      GOTO 17010
18000 VDU4:CLS:RETURN
19000 R=R+1:IF R<23 THEN RETURN
19010 PRINT TAB(0,24);:INPUT "Press RETURN to continue" I$
19020 CLS: R=0: RETURN
20000 MODE4:F9=2:VDU24,80;40;1199;864;:VDU28,0,3,39,0:VDU4:
      VDU23,240,0,0,0,0,0,224,224,224
20020 RETURN
21000 ON F9 GOSUB 20000,20020,20020,20000
21007 IF XX<NX THEN X1=XX:XX=NX: NX=X1
21008 IF XY<NY THEN X1=XY: XY=NY: NY=X1
21010 R0=XY-NY:R1=XX-NX:SX=10^(INT(LOG(R1))):
      SY=10^(INT(LOG(R0)))
21020 IF R1/SX<2 THEN SX=SX/2: GOTO 21020
21025 IF R1/SX>6 THEN SX=SX*2: GOTO 21025
21030 IF R0/SY<3 THEN SY=SY/2: GOTO 21030
21035 IF R0/SY>8 THEN SY=SY*2: GOTO 21035
21040 R0=R0/206: R1=R1/280: F9=3: RETURN
22000 U%=@%:@%=&4
22010 ON F9 GOSUB 21000,21000,20020,21000
22020 U1=1-(XX*NX>0)-2*(XY*NY>0):ON U1 GOTO 22030,22040,22050,
      22060
22030 U3=0:V3=0:MOVE FNX0(0),FNY0(NY):DRAW FNX0(0),FNY0(XY):
      MOVE FNX0(NX),FNY0(0): DRAW FNX0(XX),FNY0(0):GOTO22070
22040 U3=NX:V3=0:MOVE FNX0(NX),FNY0(0):DRAW FNX0(XX),FNY0(0):
      GOTO 22070
22050 U3=0:V3=NY:MOVE FNX0(0),FNY0(NY):DRAW FNX0(0),FNY0(XY):
      GOTO 22070
22060 U3=NX:V3=NY
22070 FOR U2=INT(NX/SX)*SX TO (INT(XX/SX)-1)*SX STEP SX:
      MOVEFNX0(U2),FNY0(V3+8*R0)
22073  VDU5:PRINTCHR$(240);LEFT$(STR$(INT(U2*1E4+.5)/1E4),5);:
      NEXT U2
22075 FOR V2= INT(NY/SY)*SY TO (INT(XY/SY)-1)*SY STEP SY
22077  IF V2<>V3 THEN VDU5:MOVEFNX0(U3),FNY0(V2+7*R0):
      PRINTCHR$(240);LEFT$(STR$(INT(V2*1E4+.5)/1E4),5);
22080  NEXT V2:@%=U%:VDU4:F9=4:RETURN
23000 MOVE FNX0(X),FNY0(Y):VDU5: PRINTA$;:VDU4:RETURN
24000 VDU4:PRINTA$:RETURN
25000 MOVEFNX0(X),FNY0(Y):DRAWFNX0(X),FNY0(Y):RETURN
26000 F8=3:IF X<NX OR X>XX OR Y<NY OR Y>XY THEN F8=0:X8=X:Y8=Y
26020 ON F8+FG+1 GOTO 26050,26050,26040,26060,26030,26040
26030 MOVEFNX0(X),FNY0(Y)
26040 DRAWFNX0(X),FNY0(Y)
26050 FG=1+F8/3:RETURN
26060 MOVEFNX0(X),FNY0(Y):FG=2:RETURN
27000 GCOL 4,1: IF X=XA OR Y=YA THEN MOVE FNX0(X),FNY0(Y):
      DRAW FNX0(XA), FNY0(YA):GOTO 27020
27010 GCOL 4,1:Y2=FNY0(Y):Y2=INT(Y2/4)*4:FOR Y1=Y2 TO FNY0(YA)
      STEP 4*SGN(YA-Y):MOVEFNX0(X),Y1:DRAWFNX0(XA),Y1:NEXT Y1
27020 GCOL0,1:RETURN
28000 IF F9=1 THEN RETURN
28010 CLG:RETURN
29000 IF F7=2 THEN RETURN
29005 IF F9=1 THEN PRINT TAB(39,24);
29010 *SDUMP
29020 RETURN
29500 GOTO29000
30000 DEF FNX(T)=EVAL(XFN$)
30001 DEF FNY(T)=EVAL(YFN$)
30002 DEF FNZ(T)=EVAL(ZFN$)
31000 DEF FNX0(X)=(X-NX)/(XX-NX)*1120+80
31010 DEF FNY0(Y)=(Y-NY)/(XY-NY)*824+40
```

B.4   USE AND LISTING OF PROTO FOR THE IBM PERSONAL COMPUTER

System requirements

The system is assumed to have IBM advanced BASIC, entitled 'BASICA' available on disc, and this should be loaded from the appropriate disc before loading any programs. This is done by typing 'BASICA' if on the default disc, or for example 'A:BASICA' if the file is on disc A.

It is also assumed that an IBM or equivalent graphics card is mounted.

To load self-contained programs

Type LOAD in capitals or lower case letters, followed by the program name within quotes; e.g.

LOAD "GRAPH" or load "graph" if on the default disc; or

load "A:GRAPH" etc if on disc A.

This method applies to all listings in the disc directories except for those ending in the letter 'X', and the program "IEM4T". In all cases except the subroutine "MATINV", the program "PROTO" needs to be merged with the main program to provide the inout and output routines described above.

To merge two programs

The program to be merged must first have been saved in the form of an ASCII rather than a normal tokenized file. If this has not been done, then the program should be loaded in the usual way, and then a version of it should be saved, usually under another name, as an ASCII file by saving with the ,A option. Thus, for example, to save 'GRAPH' back to the default drive as 'GRAPHEX', type

save "GRAPHEX",A

Programs which have been saved in this form on the IBM disc available with this book have file names ending in 'X'. Thus the appropriate file for the 'PROTO' file is called 'PROTEX'. To combine this, or another suitable program, into another, first load the first program in the usual way, e.g.

load "STORFLO"

and then type, for example

merge "PROTEX"

The two programs are then combined, with any duplicated line numbers being over-written by the merged program. The combined program can then be run, listed or saved in the normal way.

Further details of these methods are given in the DOS and BASIC manuals for the IBM Personal Computer.

Appendix B4:   Figure 1.   Listing for 'PROTO' for the IBM PC

```
10      C$="Enter optional function (see GOSUB 13000) or
        return:A$=":A$="":C=999:R=999:GOSUB 11000:IF A$="" THEN
        STOP ELSE GOSUB 13000
10000 REM Initialise machine:Setup constants:Open text screen
10010 SCREEN 0,1:CLS:F9=1:NX=0:XX=1:NY=0:XY=1:C=999:R=999:
        LG=1/LOG(10)
10020 DIM RR(20),RJ(20)
10030 A=0:A$="":C$="":RANDOMIZE TIMER
10070 GOTO 13500
11000 A0$=A$:L1=LEN(A0$):GOTO 12020:REM Input A$, cued with C$
        and prompted by current A$ at C,R
12000 A0$=STR$(A):L1=LEN(A0$) :REM Input A,cued with C$ and
        prompted by current value of A at C,R
12010 IF LEFT$(A0$,1)=" " THEN A0$=MID$(A0$,2):L1=L1-1:
        GOTO 12010
12015 GOSUB 12020:A=VAL(A$):RETURN
12020 FOR I1=1 TO L1:A0$=A0$+CHR$(29):NEXT I1:A0$=C$+A0$
12030 A5$=A0$:GOSUB 15150:PRINT A0$;:INPUT "",R$:X1=POS(0):
        Y1=CSRLIN:L2=1
12033 IF R$="" THEN RETURN
12035 A$=R$:RETURN
12500 C$="Enter "+C$+" ("+STR$(LL)+"-"+STR$(UL)+"):"
12510 GOSUB 12000:IF A<LL OR A>UL THEN GOTO 12510
12520 RETURN
13000 X1%=ASC(MID$(A$,1))-87: IF X1%>32 THEN X1%=X1%-32 :
        REM Write in a function of t, entered in form X-
        Z(t)="...":Assign its value to FNX-FNZ(t)
13005 IF X1%<1 OR X1%>3 THEN PRINT "Functions accepted only
        for X(t) to Z(t)":RETURN
13010 A1$=MID$(A$,INSTR(A$,"("))
13016 OPEN "f.bas" FOR OUTPUT AS £1 : PRINT £1,"10"
13017 PRINT £1,13500+X1%;" def fn";CHR$(X1%+87);A1$
13020 CLOSE £1:MERGE "f.bas"
13500 REM
13999 RETURN
14000 A=RND(1):RETURN:REM A=Random No in (0,1)
14100 A=LOG(A); RETURN:REM A=Natural LOG(A)
14200 A=LOG(A)*LG: RETURN:REM A=Logarithm A,base 10
15000 WIDTH 80:F9=1:RETURN:REM Open 80x24 text screen
15110 IF C=999 AND R=999 THEN C1=POS(0):R1=CSRLIN:GOTO 15130
15115 C1=C+C*(C<0)+(C-C2)*(C>C2)
15120 R2=24+23*(F9>1):R1=R+(R<0)+(R-R2)*(R>R2)
15130 LOCATE R1,C1:RETURN
15150 C2=40-LEN(A5$):C2=C2+(C2=40)+C2*(C2<0):GOTO 15110
15200 A$=MID$(B$,M,1):RETURN:REM M'th character of B$
15300 A$=MID$(B$,M,N):RETURN:REM N characters of B$, starting
        with M'th
15400 A$=LEFT$(B$,N):RETURN:REM left N characters of B$
15500 A$=RIGHT$(B$,N):RETURN:REM right N characters of B$
16000 A5$=A$:GOSUB 15150:PRINT A$;:RETURN:REM Print A$ @ Row
        R, Col C
17000 A5=A:GOSUB 15110:PRINT A:RETURN:REM Print A @ Row R,Col C
17200 FF$=STRING$(B,"£")+"."+STRING$(D,"£"): A5=A: GOSUB
```

```
15110: PRINT USING FF$;A: RETURN:REM Print fixed format
       B.D
17300 FF$=STRING$(B,"£"):A5=A:GOSUB 15110:PRINT USING
      FF$;A:RETURN:REM Print B integer format
17400 B=B+1:FF$=STRING$(B,"£")+"."+STRING$(D,"£")+"^^^^":
      A5=A: GOSUB 15110: PRINT USING FF$;A: RETURN:REM Print
      in Sci format B.D exp +**
17500 L=CINT(A):IF L=0 THEN SET L=1:
17510 NN$=STR$(L):V=LEN(NN$):V=V-1:D=B-V:FF$=STRING$(V,"£")
      +"."+STRING$(D,"£"):
17520 A5=A:GOSUB 15110:PRINT USING FF$;A:RETURN:REM Print in B
      digit free format
18000 CLS:RETURN:REM Clear screen
19000 R=R+1:IF R<23 THEN RETURN:REM Increment row r and clear
      screen when full
19010 LOCATE 23,1:INPUT "Press Return To Continue", I$
19020 CLS:R=0:RETURN
20000 SCREEN 2:CLS:KEY OFF: SROW=0:  SCO$= STRING$(79," "):
      REM High-Res graphics screen
20005 NX=0:XX=1:NY=0:XY=1:SX=.25:SY=.25:FG=1
20006 DEF FNXO(X)=75+(X-NX)/(XX-NX)*500
20007 DEF FNYO(Y)=175-(Y-NY)/(XY-NY)*140
20008 F9=2
20010 SROW=SROW+1+(SROW=2)*2:LOCATE SROW,1:PRINT SCO$;: LOCATE
      SROW,1: RETURN
21000 IF F9=1 OR F9=4 THEN GOSUB 20000:REM Set range of
      plotted values
21010 SX=(XX-NX)/4:SY=(XY-NY)/4
21013 X=(XX-NX)*2/3:Y=(XY-NY)*2/3
21020 F9=3:GOTO 20010
22000 IF F9<>3 THEN GOSUB 21000:REM Draw axis scaled at
      intervals SX,SY
22010 PSET(75,35):DRAW "d140r500l500u140"
22015 AX=CINT((SX*500)/(XX-NX)):AY=CINT((SY*140)/(XY-NY))
22020 PTSX=CINT((XX-NX)/SX):PTSY=CINT((XY-NY)/SY):PSET(75,175)
22030 FOR I=1 TO PTSY:DRAW "U=AY;":DRAW "r4l8r4":NEXT I:
      PSET(75,175)
22040 FOR I=1 TO PTSX:DRAW "R=AX;":DRAW "u4d8u4":NEXT I
22045 GOSUB 30000:F9=4:GOTO 20010
22050 RETURN
23000 REM Print A$ at X,Y on graphics screen
23005 AB=FNYO(Y)/8:AC=FNXO(X)/8
23010 LOCATE AB,AC:PRINT A$
23020 GOTO 20010
24000 PRINT A$;: GOTO 20010:REM Print A$ on text area above
      graphics screen
24500 PRINT A;:GOTO 20010:REM Print a on text area above
      graphics screen
25000 REM plot dot at X,Y
25010 PSET(FNXO(X),FNYO(Y)):GOTO 20010
26000 REM Draw from last point to X,Y;FG should be set to zero
      ·to begin new graph
26010 IF FG=0 GOTO 26030
26020 LINE -(FNXO(X),FNYO(Y)):GOTO 20010
26030 PSET(FNXO(X),FNYO(Y)):FG=1: GOSUB 20010
```

```
27000 REM Plot and fill rectangle X-XA, Y-YA
27005 FOR Z=Y TO YA STEP (XY-NY)/140
27010 LINE (FNXO(X),FNYO(Z))-(FNXO(XA),FNYO(Z)),,,&HAAAA
27015 NEXT Z
27020 GOTO 20010
28000 IF F9=1 THEN RETURN:REM Clear graphics screen
28010 CLS:RETURN
30000 REM Label graphs with X,Y values
30010 FOR I=1 TO PTSY+1:RR(I)=0:RJ(I)=NY:NEXT I:RR(1)=177
30020 FOR I=2 TO PTSY+1:RR(I)=RR(I-1)-AY:RJ(I)=RJ(I-1)+SY:
      NEXT I
30030 FOR I=1 TO PTSY+1:PSET(75,RR(I)):A=RR(I)*40/320:
      B=(75*25/200)-5:
30040 LOCATE A,B:PRINT USING "£££.££";RJ(I):NEXT I:GOTO 30050
30050 FOR I=1 TO PTSX+1:RR(I)=0:RJ(I)=NX:NEXT I:RR(1)=40
30060 FOR I=2 TO PTSX+1:RR(I)=RR(I-1)+AX:RJ(I)=RJ(I-1)+SX:NEXT I
30070 FOR I=1 TO PTSX+1:PSET(RR(I),175):A=(170*40/320)+2:
      B=(RR(I)*25/200)+1:
30080 LOCATE A,B:PRINT USING "£££.££";RJ(I):NEXT I:RETURN
```

Index

AVAILABLE

The programs described in this book are available on disk for your IBM PC (and most compatibles) and the BBC/B and Master computers.

Order the Program Disk **today** priced £14.95 including VAT/$29.60 from your computer store, bookseller, or by using the order form below.

Kirkby, Naden, Burt and Butcher: **Computer Simulation in Physical Geography** — Program Disk

Please send me .... copies of the Kirkby, Naden, Burt and Butcher: *Computer Simulation in Physical Geography*— Program Disk at £14.95 including VAT/$29.60 each

IBM PC   0 471 91836 9   BBC/B and Master   0 471 91837 7

POSTAGE AND HANDLING FREE FOR CASH WITH ORDER OR PAYMENT BY CREDIT CARD

☐ Remittance enclosed .... Allow approx. 14 days for delivery

☐ Please charge this to my credit card (All orders subject to credit approval)

Delete as necessary: — AMERICAN EXPRESS, DINERS CLUB, BARCLAYCARD/VISA, ACCESS

CARD NUMBER   ⬚⬚⬚⬚⬚⬚⬚⬚⬚⬚⬚⬚⬚⬚   Expiration date ....

☐ Please send me an invoice for prepayment. A small postage and handling charge will be made.

Software purchased for professional purposes is generally recognized as tax deductible.

☐ Please keep me informed of new books in my subject area which is ....

NAME/ADDRESS ....

....

....

OFFICIAL ORDER No ....   SIGNATURE ....

If you have any queries about the compatability of your hardware configuration, please contact:

Helen Ramsey
John Wiley & Sons Limited
Baffins Lane
Chichester
West Sussex
PO19 1UD
England

Customer Service Department
John Wiley & Sons Limited
Shripney Road
Bognor Regis
West Sussex
PO22 9SA